Cambridge Ele

Elements in the Philosophy of Biology
edited by
Grant Ramsey
KU Leuven
Michael Ruse
Florida State University

PHILOSOPHY OF IMMUNOLOGY

Thomas Pradeu
CNRS & University of Bordeaux

CAMBRIDGE
UNIVERSITY PRESS

University Printing House, Cambridge CB2 8BS, United Kingdom

One Liberty Plaza, 20th Floor, New York, NY 10006, USA

477 Williamstown Road, Port Melbourne, VIC 3207, Australia

314–321, 3rd Floor, Plot 3, Splendor Forum, Jasola District Centre, New Delhi – 110025, India

79 Anson Road, #06–04/06, Singapore 079906

Cambridge University Press is part of the University of Cambridge.

It furthers the University's mission by disseminating knowledge in the pursuit of education, learning, and research at the highest international levels of excellence.

www.cambridge.org
Information on this title: www.cambridge.org/9781108727501
DOI: 10.1017/9781108616706

First published 2019

A catalogue record for this publication is available from the British Library.

ISBN 978-1-108-72750-1 Paperback
ISSN 2515-1126 (online)
ISSN 2515-1118 (print)

Philosophy of Immunology

Elements in the Philosophy of Biology

DOI: 10.1017/9781108616706
First published online: November 2019

Thomas Pradeu[1,2]

Abstract: Immunology is central to contemporary biology and medicine, but it also provides novel philosophical insights. Its most significant contribution to philosophy concerns the understanding of biological individuality: what a biological individual is, what makes it unique, how its boundaries are established, and what ensures its identity through time. Immunology also offers answers to some of the most interesting philosophical questions. What is the definition of life? How are bodily systems delineated? How do the mind and the body interact? In this Element, Thomas Pradeu considers the ways in which immunology can shed light on these and other important philosophical issues. This title is also available as Open Access on Cambridge Core.

Keywords: immune system, self, individuality, cancer, neuroimmunology

ISBNs: 9781108727501 (PB), 9781108616706 (OC)
ISSNs: 2515-1126 (online), 2515-1118 (print)

---

1. ImmunoConcept (UMR5164), CNRS and University of Bordeaux, France.
2. Institut d'histoire et de philosophie des sciences et des techniques (UMR8590), CNRS and Panthéon-Sorbonne University, France.

Contents

1 Introduction: The Centrality of Immunity in Biology and Medicine 1

2 Immunity: A Matter of Defense? 3

3 The Unity of the Individual: Self–Nonself, Autoimmunity, Tolerance, and Symbiosis 13

4 Cancer as a Deunification of the Individual 29

5 Neuroimmunology: The Intimate Dialogue between the Nervous System and the Immune System 43

References 64

1 Introduction: The Centrality of Immunity in Biology and Medicine

Immunology is one of the most central and dynamic fields of today's biological and biomedical sciences. It constitutes, in fact, a pivotal bridge between basic biology and medicine. Immunology is generally defined as the domain studying the defense of the organism against pathogens but its scope is actually much wider. Topics as diverse as cancer, infectious diseases, vaccination, transplantation, autoimmune diseases, chronic inflammatory diseases, metabolic diseases, development, aging, repair and regeneration, and host–microbiota interactions, among many others, are all directly related to the field of immunology. Furthermore, it now appears that immune systems exist almost ubiquitously across the living world (including in animals, plants, and prokaryotes). In fact, virtually all domains of biology and medicine are connected to immunology, and when opening recent issues of leading science or medicine journals, one can get the impression that immunology is omnipresent. Why has immunology become so central in our science and daily lives – and why does this matter philosophically?

I work as a philosopher of science embedded in an immunology lab affiliated with the Bordeaux University hospital. Over the years I have become increasingly aware of the key role played by the immune system in practically all kinds of diseases – in their aetiology, diagnosis, and treatment. If you receive a transplant, the biggest challenge is immunological rejection of the graft, which explains why you will be prescribed immunosuppressive drugs. If you have cancer, depending on the type of tumor, you might receive one of the now hugely discussed immunotherapies, an advance that was awarded the 2018 Nobel Prize in Physiology/Medicine and which constitutes an immense hope for medical doctors and patients worldwide (Ribas and Wolchok 2018). Even if you do not receive immunotherapies, the number of the different populations of your immune cells will be checked regularly to choose and adapt your treatment. Immunology is also central, naturally, for our understanding of autoimmune diseases: in type 1 diabetes, for instance, immunologists seek to explain why the immune system selectively destroys pancreatic β-cells (which secrete insulin), and how this process might be controlled (Lehuen et al. 2010). If you happen to come back from another country with a bad viral or bacterial infection, again, the main goal of medical doctors will be to make your immune system cope with that infectious agent without severely disturbing the balance of your immune responses to other elements. Moreover, vaccination rests on the idea of stimulating the immune system against a particular target. The immune system also plays a major role in pathologies as diverse as cardiovascular diseases (Hansson and Hermansson 2011), neurodegenerative disorders (Heneka et al. 2014; Heppner et al. 2015), and obesity (Lumeng and Saltiel 2011) – and for all these diseases it constitutes one important point of leverage used in the clinic.

So, immunology is pretty much unavoidable in our daily lives, both in health and disease. One aim of this Element is to show that immunology is also of paramount importance for philosophers. The most central contribution of immunology to philosophy concerns, arguably, the understanding of biological individuality. From the end of the nineteenth century onward, it has been recognized that immunology raises key questions about what a biological individual is, what makes it unique, how its boundaries are established, and what ensures its identity through time (Tauber 1994; Pradeu 2012). This Element will explore other philosophical lessons that can be drawn from current immunology – including the definition of life (or, more specifically, the basic requirements for all living things), the delineation and regulation of bodily systems, part–whole relations, the notion of biological function, and mind–body interactions.

The main claims made in this Element are summed up in Box 1.1.

The present philosophical exploration of immunology will be made through the examination of concrete scientific and medical examples, such as host–microbe symbioses, cancer immunotherapies, and the CRISPR-Cas systems.

Box 1.1 Main Claims Made in the Present Element

1. Most (perhaps all) living things possess an immune system.
2. Immunity is not limited to the activity of defense. The immune system plays a central role in activities as diverse as development, tissue repair, and clearance of debris, among others.
3. Anyone interested in biological individuality must take into account what immunology says on this question.
4. The immune system plays a key role in delineating (and constantly redrawing) the boundaries of a biological individual, determining which elements can be part of that individual, and insuring its cohesion.
5. Cancer results from a process of decohesion in a multicellular organism, and the immune system has a major influence on the control of this process.
6. The nervous system and the immune system intimately interact. Neuroimmunologists' claims that the immune system can influence behavior and even cognition are worth examining.

This Element tries to talk simultaneously to philosophers, scientists, and medical doctors. To philosophers of biology, this Element says that immunology raises many crucial conceptual and philosophical issues and can integrate elements coming from several related biological fields, including microbiology, developmental biology, physiology, evolution, and ecology. For philosophers and metaphysicians, this Element argues that immunology can shed new light on some philosophical questions that have been fundamental since at least the time of Aristotle, such as what constitutes the identity of an individual through time. The message to biologists is that the immune system must be rethought as one of the most basic and indispensable aspects of any living thing. Finally, the suggestion to medical doctors is that a constant reflection on immunological concepts can help open up novel therapeutic avenues – for instance, about cancer, autoimmune diseases, or the management of ecological interactions between microbes within our bodies. Overall, the approach taken in this Element will be an example of philosophy *in* science, that is, a type of philosophical work that aims at interacting intimately with science and contributing to science itself (Laplane et al. 2019).

The contents of the Element are as follows. Section 2 critiques the idea that immunity should be defined exclusively in light of the concept of defense of the organism against external threats and extends immunity to other key dimensions, particularly development, repair, and other housekeeping activities. Section 3 shows that immunology is central to the definition of biological individuality and proposes that a physiological individual is a community of heterogeneous constituents, including microbes, unified by the action of the immune system. Section 4 examines the claim that cancer can be defined as a breakdown of biological individuality and argues that the immune system can both prevent and promote this breakdown. Finally, Section 5 explores the interactions between the nervous system and the immune system and assesses the claim that the immune system may be involved in behavior and cognition.

2 Immunity: A Matter of Defense?

If you cut yourself with unclean tools while doing some gardening, the pervasive bacterium *Staphylococcus aureus* might enter into your body via the wound site. If your immune system works normally, you will certainly get rid of these bacteria rapidly. An oversimplified description of this process is that tissue-resident cells, especially macrophages, detect the bacteria, trigger a local inflammation (which facilitates the immune response), and usually eliminate the bacteria, sometimes with the help of other innate immune cells (like neutrophils) that are recruited at the site of infection. If the bacteria are not promptly eliminated, then antigen-presenting cells, typically dendritic cells, migrate to secondary lymphoid organs such as lymph nodes and present bacterial fragments to naive lymphocytes circulating in these compartments. Lymphocytes with high specificity and affinity for these bacterial antigens are activated, and their populations expand. Specific lymphocytes then migrate to the infection site, and, in concert with many other cellular and molecular components such as antibodies, they destroy the bacteria.

The highly intertwined processes that collectively constitute an immune response suggest that our immune system is truly a system – a set of processes that involve many interacting cells distributed throughout the body. Indeed, although it comprises particular cells (Box 2.1) and organs, the immune system exerts its influence everywhere in the organism, especially via its network of lymphatic vessels and its numerous tissue-resident cells (Figure 2.1).

The system by which an organism defends itself against pathogens is precisely what has generally been called the *immune system* (Janeway 2001; Paul 2015). Is the activity of the immune system, though, only a matter of defense? In this section, I show that immunity has been understood historically as an organism's capacity to defend itself against pathogens, and that defensive immune mechanisms have been identified in all species. I then argue that the immune system cannot be reduced to its defense activity and promote on this basis an extended view of immunity. Next, I explore the complexities of accounting for the evolution of immunological processes and attributing a single function to the immune system. Finally, I explain why it is difficult in today's immunology to offer a definition of immunity.

Box 2.1 Simplified Presentation of the Main Cellular and Molecular Components of the Immune System in Mammals, with Some of Their Activities

1 Cells

Macrophages: phagocytosis, elimination of pathogens, clearance of debris, antigen presentation, tissue repair.

Neutrophils: phagocytosis, elimination of pathogens, chemotaxis, constitution of neutrophil extracellular traps, tissue repair.

Mast cells: elimination of pathogens, wound healing, immune tolerance.

Dendritic cells: antigen uptake at the periphery, antigen presentation in secondary lymphoid organs.

Natural killer cells: elimination of infected cells and cancer cells.

Innate lymphoid cells: elimination of pathogens, tumor surveillance, tissue repair, metabolism.

Effector T lymphocytes: stimulation of other immune cells, destruction of infected cells.

Regulatory T lymphocytes: downregulation of other immune cells, prevention of autoimmune diseases.

B *lymphocytes*: neutralization, opsonization (promotion of phagocytosis), complement activation.

2 Molecules

Complement: phagocytosis, inflammation, membrane attack.

Cytokines, including chemokines, interferons, and interleukins: cell signaling.

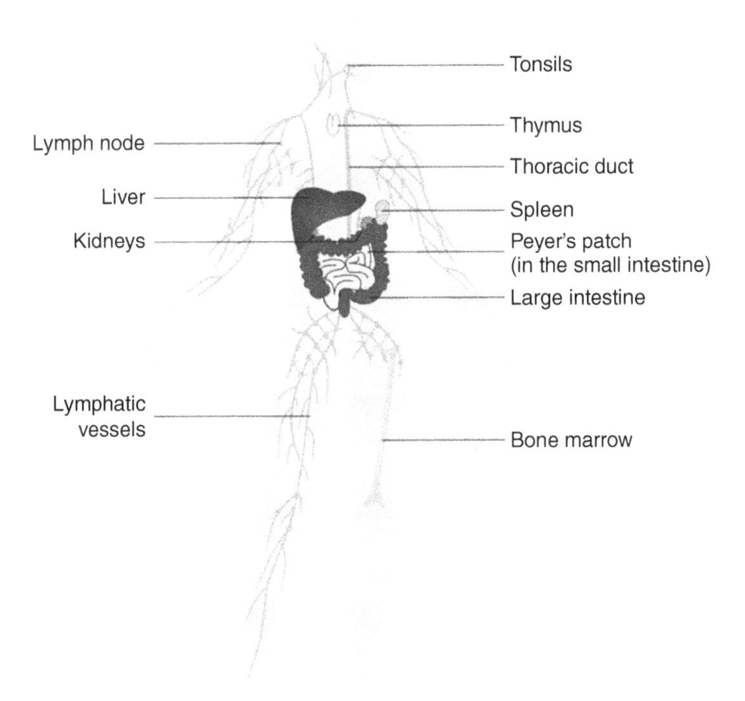

Figure 2.1 Human immune system. The human immune system, which comprises different organs (thymus, bone marrow, spleen, lymph nodes, and so on), different cells (both circulating and resident) and molecules, and a network of lymphatic vessels, exerts its influence everywhere in the organism. (Figure drawn by Wiebke Bretting).

2.1 Historically, Immunity Has Been Understood As the Capacity of an Organism to Defend Itself against Pathogens

Pathogens have constituted a major force shaping the evolution of human beings throughout their history. Devastating infectious diseases are a clear example. Plague, caused by the bacterium *Yersinia pestis*, killed more than 50 million people in the fourteenth century. More recently, the pandemic of influenza virus following the First World War killed over 40 million people worldwide.

It has long been recognized that humans can resist some infections, with important differences between individuals and/or populations. While some people are killed, others show disease symptoms but survive, and still others do not seem to be affected at all. The notion of immunity is meant to capture this idea of a specific capacity to avoid the detrimental effects of a pathogen. Immunity is generally defined as the capacity by which an organism can defend itself against pathogens.

Host–pathogen interactions are complex, as each partner adapts to the other. This emphasizes the need to always understand immune defense and pathogenicity in their ecological and evolutionary contexts (Box 2.2).

Box 2.2 Importance of the Ecological and Evolutionary Context When Discussing Immune Defense and Pathogenicity

Immune defense and pathogenicity are not intrinsic properties of hosts and microbes. Rather, they are a matter of evolutionary and ecological context. Approaches that pay attention to evolution and ecology have historically played an important role in immunology (Burnet 1940; Méthot and Alizon 2014). They have taught three important lessons.

1 Pathogenicity Is a Gradual and Contextual Phenomenon

An infectious microorganism is not intrinsically pathogenic (Casadevall and Pirofski 1999; Méthot and Alizon 2014): it can be harmful in one species and benign in another, and its virulence often varies between different individuals in a population. Even within an individual, pathogenicity depends on pathogen localization, host physiological and immunological state at the moment of the infection, the presence of other microorganisms, and the past interactions with this pathogen or others, among many other factors. In addition, the vast majority of microorganisms are not harmful to their hosts.

2 An Ever-Going "Arms Race" Occurs between Hosts and Pathogens

Hosts and invaders continuously adapt one to the other. This constitutes one of the main reasons why immune systems are so intricate, with so many different components acting at various levels. Often, a pathogen evolves a way to evade a given recognition system of its host species, but then the host species evolves new recognition systems, which in turn might be circumvented by the pathogen. Such host–pathogen competition often takes the form of manipulation of the immune system by pathogens (Finlay and McFadden 2006). For example, various bacteria display molecular patterns that look like those of the host (molecular mimicry). Furthermore, some bacteria can establish residence within immune cells, which enables them to partly escape the immune response – as *Mycobacterium tuberculosis* does in macrophages, for instance.

3 Immune Defense Comes at a Cost

As emphasized by the recent field of ecological immunology (or ecoimmunology (Schulenburg et al. 2009)), immune responses are costly because the immune system takes up many bodily resources, and if things go wrong, these responses can cause terrible damage to the organism. Ecological immunology has also shown the existence of trade-offs between the different physiological responses of a host to various environmental challenges, for example, between immunity, reproduction, growth, and thermoregulation (Schulenburg et al. 2009).

Historically, the definition of immunity as defense against pathogens is intimately connected with the development of the vaccination technique (Moulin 1991; Silverstein 2009). The birth of immunology as a biomedical field was related with the process of so-called immunization, that is, the acquisition of protection against a specific agent. Etymologically, immunity means an exemption (the right to legally escape specific taxes in ancient Rome). The development of large-scale scientific vaccinations, particularly with Robert Koch and Louis Pasteur in the nineteenth century, is often seen as one of the main foundations of the field of immunology (Bazin 2011). Vaccination is connected with the idea that an organism can increase its defense capacity by "learning" how to neutralize a given pathogen. The vaccinated organism responds quicker and more strongly if it reenters into contact with the same pathogen. This capacity is called *immunological memory*, a phenomenon that

Box 2.3 The Distinction between Innate and Adaptive Immunity and Its Limits

Immunologists often distinguish between innate and adaptive immunity: innate immunity is supposed to correspond to a quick immune response, without training, whereas adaptive immunity takes more time and entails the capacity to "learn" from previous encounters with a given target. (A more precise definition says that innate immunity is characterized by germline-encoded receptors, while adaptive immunity is characterized by the production of novel immune receptors via somatic recombination and clonal expansion (e.g., Lanier and Sun 2009).) Innate and adaptive immunity, however, intimately and dynamically interact. Furthermore, over the last two decades the distinction between them has blurred because many immune components can be located on a gradient from innate to adaptive immunity (Flajnik and Du Pasquier 2004), and immunological memory is in fact a complex, multidimensional, and gradual process found across the living world, including in bacteria and archaea (Pradeu and Du Pasquier 2018).

has long been used to make a distinction between two arms of immunity, namely innate and adaptive immunity (Box 2.3).

2.2 Defensive Immune Mechanisms Have Been Identified in Virtually All Living Things

Beyond humans, all animals, plants, unicellular eukaryotes, bacteria, and archaea are constantly under the potential threat of pathogens and have evolved multiple mechanisms to cope with those pathogens (Anderson and May 1982; Stearns and Koella 2008). Contrary to the long-held view that only vertebrates possess an immune system, in the last thirty years or so immune systems have been found in all the species in which their presence has been thoroughly investigated (Pradeu 2012). Importantly, one can observe that, in all these cases, the criterion used for saying that an immune system exists in a species has been the identification of a system of recognition, control, and elimination of pathogens. This confirms that defense remains the intrinsic, if sometimes implicit, definition of immunity that most biologists adopt when they talk about the immune system.

Plants lack a circulatory system and mobile immune cells, but they are capable of establishing immune responses that are highly specific, with limited damage to the host, and that can even generate a form of immunological memory (Spoel and Dong 2012). Plants deal with pathogens by diverse modes of recognition (including nucleotide-binding domain, leucine-rich repeat (NLR) receptors) and effector responses (Jones and Dangl 2006).

Bacteria and archaea can be infected by pathogens, including viruses called *bacteriophages*, or more simply *phages*. They respond to these pathogens through different immune mechanisms, including suppression of phage adsorption, restriction modification of the invading phage genome, abortive infection, and the recently discovered CRISPR-Cas systems (CRISPR stands for "**C**lustered, **R**egularly **I**nterspaced, **S**hort **P**alindromic **R**epeats") (Hille et al. 2018). CRISPR-Cas systems provide many bacteria and archaea with protection against phages and other mobile elements (including plasmids and transposons) through a three-step process. The first step is adaptation: small fragments of DNA from the invader are incorporated into the CRISPR array of the host. The second step is expression and processing: the CRISPR array is transcribed, and the precursor transcript is processed to generate CRISPR RNAs. The final step is interference: the CRISPR RNA guides a complex of Cas proteins to the matching target, which initiates the destruction of the invading nucleic acid (Jackson et al. 2017). CRISPR-Cas has generally been described as a key defense mechanism of prokaryotes against mobile elements (including phages) (Horvath and Barrangou 2010), along with other, more recently identified antiphage defense systems (Doron et al. 2018). Most experts even consider CRISPR-Cas as a form of *adaptive* immunity in bacteria and archaea (because it displays a form of immunological memory) on top of being heritable, as it can be transmitted to daughter cells (Horvath and Barrangou 2010; Hille et al. 2018). This has led to discussions over the potentially Lamarckian nature of the CRISPR-Cas system (see (Koonin 2019) and accompanying commentaries).

Other important examples of organisms with now well-described immune systems include insects (Lemaitre and Hoffmann 2007), sponges, hydra, and slime molds (Müller 2003; Chen et al. 2007; Augustin et al. 2010).

2.3 Extended Immunity: Immunity Goes Well Beyond Defense

The centrality of host defense against pathogens in the survival of all organisms should not obscure the fact that immunity goes well beyond mere defense. Although immunity was long conceived exclusively as a system of defense, recent research has shown that immune processes actually play a critical role in a wide variety of physiological phenomena such as development, tissue repair, clearance of debris or dead cells, maintenance of local tissue functions, metabolism, thermogenesis, and the functioning of the nervous system, among many others (Pradeu 2012; Rankin and Artis 2018), leading to an extended view of immunity. Repair is a particularly fundamental process, which maintains the integrity and cohesion of the organism.

When you cut yourself, in addition to the well-known action of platelets, which initiate blood clot and vasculature closure, a horde of immune cells is required at every stage of the repair process to insure proper wound healing (Gurtner et al. 2008; Wynn and Vannella 2016). The three key stages are inflammation, new tissue formation, and remodeling. The main immune cells participating in tissue repair are neutrophils and macrophages. The plasticity and adaptation to context of these cells are crucial (Laurent et al. 2017). Immune cells are also important in processes of regeneration, such as those found in plants, hydra, arthropods, and amphibians, and by which entire complex structures such as limbs can regrow (Eming et al. 2014). Another key daily activity of the immune system is the clearance of bodily debris, coordinated by phagocytotic cells (Nagata 2018). In addition, the immune system is essential for development, that is, the early construction of the organism. This includes the indispensable role of immune-mediated apoptosis and phagocytosis very early on in many developmental processes (Wynn et al. 2013; Okabe and Medzhitov 2016), as well as the role of the complement in development (Ricklin et al. 2010; Stephan et al. 2012) (e.g., phagocytosis mediates the indispensable elimination of excessive tissues; the complement, a cascade of proteins in the blood, remodels synaptic connections in the developing nervous system). Importantly, even though the above description applies mainly to animals, the observation that the immune system realizes various activities beyond defense holds across all living organisms; for example, CRISPR-Cas systems in prokaryotes participate not only in defense but also in repair.

Many of the same actors that ensure the defense of the organism against pathogens, therefore, are equally central for processes previously considered as nonimmune and which overlap to a significant degree (Figure 2.2). Perhaps one could even consider that the very division of these processes into such categories as "defense," "repair," and "development" reflects more the way we, as investigators, address questions about bodily systems (and divide such processes into convenient categories) than real differences in nature. From this point of view, there would be much to say in favor of a revised epistemology of immunology, understood as a reflection on how the categories by which the immune system has been conceived (in the context of the division into systems) could be redefined and redrawn.

2.4 Accounting for the Evolution of Immunological Processes and Attributing a Function to the Immune System Have Become Difficult

The extension of immunity well beyond defense has consequences for two central issues, namely the evolutionary history of immune systems and

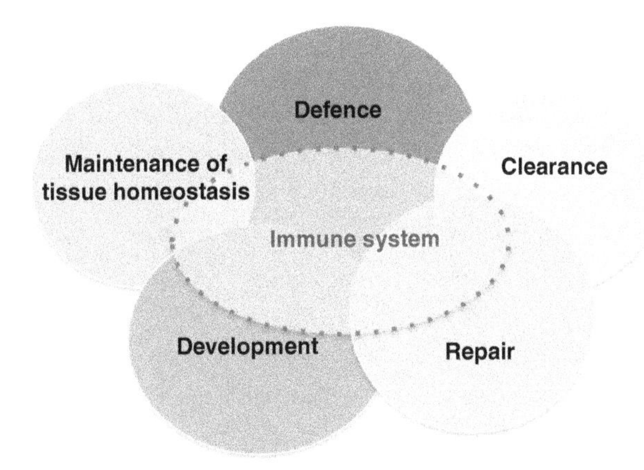

Figure 2.2 Extended immunity: overview of the various, partly overlapping, activities of the immune system. The immune system not only defends the organism against different potential threats but also constructs, repairs, cleans up the organism, and maintains tissue homeostasis, among other activities.

functional ascription in the case of the immune system. It increasingly appears that immune systems have evolved under different selective pressures, which include the necessity to defend the organism against pathogens but are much more diverse than that. Natural selection has favored immune systems efficient at simultaneously defending, constructing, repairing, cleaning up, and maintaining the organism. Many trade-offs exist among these various selective pressures. For example, so-called type 2 immunity (essential for repair, but also in response to parasites) is partly at odds with type 1 immunity (which responds to intracellular abnormalities, including viruses and intracellular bacteria), as the activation of one weakens the efficacy of the other. Current immune systems function by balancing these different dimensions and reaching a complex equilibrium between them in the context of present and past environmental pressures (Eberl 2016).

This discussion also brings up the question of the origins of the first immune systems. The capacity to regulate the various processes of defense, development, repair, clearance of debris, and maintenance of homeostasis is probably as ancient as life itself. If so, then all living things that currently exist or have existed in the past have an immune system. The key question, therefore, becomes how replicating molecules started to acquire, a few billions of years ago, the capacity to coalesce into groups, stick together, and maintain, repair, and defend themselves at this emerging group level. Much later, the same question was raised for the transition from unicellular to multicellular organisms where, again, the constitution of an immune system at the multicellular level was crucial (Michod 1999; Pradeu 2013). Obviously, this approach to the evolution of immune systems is entirely different from the classic (and naturally important) discussion in immunology over how adaptive immune systems evolved in vertebrates some 400 to 500 million years ago (Pancer and Cooper 2006). Much work remains to be done on this question of the origins of immunity, including by developing cross-species comparative immunology.

What has been said here about the expansion of the scope of immunity can also be framed in terms of the long-standing debate in the philosophy of biology about functions (Wright 1973; Cummins 1975). Bodily systems generally have been viewed as obvious instances of a valid functional ascription: the function of the digestive system is to digest, the function of the respiratory system is to breathe, and so on. According to the traditional view of immunity, the function of the immune system is to defend the organism. Following this tradition, some philosophers have suggested (Matthen and Levy 1984; Melander 1993) that the function of the immune system, in the etiological sense (that is, what it has been selected for) (Wright 1973), is organism defense. Yet saying what the immune system has been selected for proves, in fact, extremely difficult. Given the various activities of the immune system and the diversity of selective pressures it has undergone, it would be inadequate, or at least too restricted, to say that the immune system has been selected for its capacity to defend the organism against pathogens. It is not possible to single out one of these aspects and say that this is "the function" (or even "the main function") of the immune system. It seems likely that the systemic approach to functions (Cummins 1975) is certainly better suited for immunology, but a detailed analysis of the promises and limitations of a systemic approach to immune functions remains to be done.

2.5 Is It Still Possible to Offer a Precise and Simple Definition of Immunity?

The extension of the scope of immunity well beyond defense affects dramatically the meaning of the term immunity and raises a thorny but central problem: given the diversity of phenomena in which the immune system plays a crucial role, is there still a unity of the concept of immunity, or is it just the loose and artificial putting together of a horde of processes under a single term? In other words, is there ultimately such a thing as immunity that unifies all of these processes under a single concept? I see this question as one of the most fascinating that current and future immunology must face. Defining what counts as an immune process and delineating the field of immunology has indeed become extremely challenging nowadays. Focusing on defense against pathogens would be too narrow. On the other hand, extending the definition of immunity so far as the overall physiological regulation of the body (as sometimes suggested by recent studies on the role of the immune system in metabolism, tissue repair, homeostasis, development, and so on) would be so broad that it might cease to be scientifically fruitful, as almost everything in biology could be said to be immunological, at least to some degree. One possibility would be to say that any component involved in a series of activities that *includes* defense against pathogens is part of immunity (in this conception, immunity would be centered on, but not limited to defense), but this option would still leave us with a broad conception of immunity. Uncertainties of this kind have led to a situation in which, perhaps unsurprisingly, current immunologists do not offer a single and consensual response to the question "What is immunity?"

I have tried in the past to put forward a definition of immunity according to which immunological processes correspond to all the processes of biochemical recognition of a target, followed by the elimination or the regulation of the elimination of this target (Pradeu 2012). This definition is not without difficulties (particularly concerning the level(s) at which it is supposed to apply), but I think it does capture the diversity of phenomena currently studied by immunologists, while remaining sufficiently specific so as not to include all classically understood physiological processes occurring in the organism. It also has the advantage of reflecting the diversity of immune systems in nature, from prokaryotes to vertebrates. More research is needed, though, to determine whether a definition of this kind could satisfy the majority of immunologists.

A related but broader and more imprecise characterization of immunity is to say that immunity is one of the main devices insuring the cohesion of the organism and the delineation of its boundaries. We will explore and assess this view in the next section, which is devoted to immunology's contribution to the definition of biological individuality.

In summary, this section has shown that immune defense is essential in every living thing, but immunity cannot be reduced to defense.

3 The Unity of the Individual: Self–Nonself, Autoimmunity, Tolerance, and Symbiosis

In 1998, Clint Hallam, a patient from New Zealand who was operated on in Lyon, France, received the world's first hand graft (Dubernard et al. 1999). The operation was a technical success, and initially the recipient seemed to feel all right. Yet he soon started to consider as "other" (foreign) the transplanted hand, which became unbearable for him. Hallam interrupted immunosuppressive treatment, organ rejection started, and eventually he asked that his hand be amputated (Dubernard et al. 2001). With face transplants (36 patients underwent face transplants from 2006 to 2016 (Siemionow 2016)), recipients and doctors are confronted with even more difficult and pressing identity issues (Carosella and Pradeu 2006).

These are certainly extreme cases. Nonetheless, every year in the United States approximately 16,000 kidney and almost 6,000 liver transplantations are done (Wolfe et al. 2010), and in the European Union around 19,000 kidney and 7,000 liver transplants (European Directorate for the Quality of Medicines & HealthCare 2015). Despite the apparently routine nature of such transplantations, the body does not seem ready to tolerate any component coming from another organism. First, in the case of the kidney, for example, medical doctors, before proceeding to transplantation, carefully check compatibility between donor and recipient, particularly in terms of the major histocompatibility complex (MHC) (Kahan 2003). Second, the transplant patient receives immunosuppressive drugs. Thus, even if not all these transplantations prompt the kind of identity crisis experienced by Clint Hallam, they do raise the questions of how the boundaries of a biological individual are defined and under which conditions some external elements can become part of that biological individual. These are exactly the questions that led the nascent field of immunology, in the first half of the twentieth century, to take on the problem of biological individuality via a combination of reflections about transplantation and infectious diseases.

The aim of this section is to show why immunology makes a critical contribution to the problem of biological individuality, especially from the points of view of boundaries and parthood just mentioned. Immunology has raised this problem mainly through the concepts of "self" and "nonself," so much so that in the 1980s immunology was famously named "the science of self-nonself discrimination" (Klein 1982). I examine here the conceptual and experimental limitations of the self–nonself framework, while insisting that this critique does not undermine the essential claim that immunology plays a key role in the definition of biological individuality.

The outline of the section is as follows. First, I describe the birth of the self–nonself framework in immunology. I then spell out how current immunologists conceive biological individuality, especially on the basis of recent research on autoimmunity, immunological tolerance, and symbiosis. Third, I explore three arguments demonstrating the central contribution of immunology to biological individuality. Fourth, I show that the functioning of the immune system sheds light on how a set of heterogeneous constituents can be turned into an integrated individual. Finally, I make a claim for combining immunology's lessons about biological individuality with the lessons drawn from other biological fields.

3.1 From Early Reflections about Immunological Individuality to the Concepts of "Self" and "Nonself"

The key contribution of immunology to the reflection on biological individuality and identity ever since the end of the nineteenth century has been emphasized by historians and philosophers of this field over the last three decades (Löwy 1991; Moulin 1991; Tauber 1994; Pradeu 2012; Anderson and Mackay 2014). Here my goal is not to go into the details of that history, but rather to show how these early thoughts have been revisited in recent immunology.

The idea that immunological reactions had something to do with the questions of bodily individuality and identity was expressed as early as the end of the nineteenth century by many scientists, including, for example, Charles Richet (Richet 1894; Löwy 1991). The many attempts to transplant body parts (especially the skin, which happens to be, in fact, an extremely difficult transplantation) in the wake of the First World War, however, constituted an important turning point, as it led many biologists and medical doctors to raise specifically the question of the biological determinants of the uniqueness of each individual and its capacity to recognize these unique determinants and to potentially detect and respond to elements that differ from them (Loeb 1937; Medawar 1957).

Starting from the 1940s, Frank Macfarlane Burnet (1899–1985) (Burnet 1940; Burnet and Fenner 1949) framed the concepts of immunological "self" and "notself" (later called "nonself") (Löwy 1991; Moulin 1991; Tauber 1991; Anderson and Mackay 2014). In its most basic form, the self–nonself framework says that the elements originating from an organism (the "self") do not trigger an immune response, while elements foreign to this organism (the "nonself") trigger an immune response. Such a framework can account for the rejection by the body of both infectious agents and grafts. The self–nonself vocabulary played an important role in establishing the study of infectious diseases as well as research on transplantation as subdomains of the science of immunology.

Burnet saw as a major challenge the question of how an organism can *learn* to preserve the constituents of the "self." Partly inspired by Ray Owen's work on chimerism (Owen 1945), Burnet postulated that, if foreign material was implanted early in the embryo, no antibodies would develop against that particular foreign material (Burnet and Fenner 1949). This was later confirmed by experiments made by Peter Medawar's group, showing that a tissue implanted early in the mouse embryo could subsequently be tolerated

(Billingham et al. 1953). For Burnet, therefore, the self is acquired, not innate: the organism acquires at an early ontogenetic stage the capacity to recognize its own constituents and to avoid their destruction (Burnet and Fenner 1949; Burnet 1969). But Burnet argues that, in nature, contrary to what happens in these experimental settings, the immunological self reflects the genetically endogenous components of the organism because the repertoire of immune receptors is constituted on the basis of these endogenous elements present in the body (Burnet 1962). Burnet considers several obvious challenges to the self–nonself framework, including autoimmune diseases and foeto-maternal tolerance as well as other forms of immunological tolerance (that is, the absence of destruction of a foreign entity by the immune system). But he treats them as *exceptions* to the general rule and considers that these exceptions must be explained via specific mechanisms (pathological mechanisms in the case of autoimmune diseases, and particular provisory mechanisms in the case of foeto-maternal tolerance) (Burnet 1969). All these reflections constituted the foundations of a fruitful theoretical and conceptual framework about the self and nonself, which was developed by Burnet over more than three decades, and which combined in an innovative way molecular, cellular, ecological, and medical considerations (Burnet 1940, 1969, pp. 309–310). Burnet shared the 1960 Nobel Prize in Physiology/Medicine with Medawar for what was essentially a contribution of a conceptual and hypothetical nature, as acknowledged by Burnet himself (Burnet 1960, p. 700).

From a theoretical viewpoint, what Burnet sees as his main contribution to immunology is not primarily the self–nonself theory but rather the "clonal selection theory." The clonal selection theory goes against instructionist approaches to antibody formation (especially that of Linus Pauling) by stating that, when an antigen penetrates into the organism, immune cells bearing receptors specific for this antigen undergo selection and are subsequently responsible for the elimination of that antigen (Burnet 1959, p. 54). Burnet therefore proposes the adoption of a Darwinian framework at the cell level within the multicellular organism (Burnet 1959, p. 64) – an idea that had been suggested before him but that Burnet framed in a much more precise way (Schaffner 1992; Silverstein 2002).

Nonetheless, the clonal selection theory and the self–nonself theory are intimately connected, so much so that Burnet generally argues for them in parallel (e.g., (Burnet 1959, 1969)). The phenomena that the clonal selection theory seeks to explain are immune recognition of the antigen, immune tolerance, and the acquisition of "self-knowledge" by the organism. To account for these phenomena, Burnet proposes the existence of two selective processes: one occurs in adult life, when an antigen enters into the body, as we just saw; but another selective process occurs long before, in early life, and this other process is the elimination of immune cells that recognize "self" components (Burnet 1959). This explains why, in 1969, Burnet writes that the clonal selection theory provides "the simplest possible interpretation of how the body's own constituents are shielded from immunological attack" (Burnet 1969, p. 12).

From the second half of the twentieth century to the present day, the self–nonself framework has been adopted by a large majority of immunologists and enriched by a host of experimental, conceptual, and theoretical contributions (e.g., Bretscher and Cohn 1970; von Boehmer and Kisielow 1990; Janeway 1992; Langman and Cohn 2000). Today, the self–nonself remains the dominant, if often implicit, framework in which immunologists conceive how the immune system works (e.g., Stefanová et al. 2002; Goodnow et al. 2005; Jiang and Chess 2009; Fulton et al. 2015). This is particularly illustrated by the fact that when a novel immune system is identified, as recently happened with the CRISPR-Cas systems in archaea and bacteria, scientists spontaneously use the self–nonself vocabulary to account for its functioning (Nuñez et al. 2015) (this comes in addition to the idea of CRISPR-Cas as a system of *defense*, as discussed in the previous section). Together with the persistent use of the self–nonself vocabulary in several other domains (including, for instance, studies on autoimmune diseases and transplantation), this confirms the long-standing influence of this framework in immunology over the last six decades.

3.2 Autoimmunity, Tolerance, and Symbiotic Interactions with Microbes

Despite its undeniable success as an encompassing conceptual framework for immunology, the self–nonself theory faces many difficulties. First, far from being always pathological, autoimmunity has been proved to be a necessary component of everyday immunity. A degree of autoreactivity (i.e., reaction to "self") characterizes the lymphocytes generated and selected in primary lymphoid organs as well as naïve lymphocytes circulating in the periphery (Tanchot et al. 1997; Anderton and Wraith 2002; Hogquist and Jameson 2014). Effector T cells are selected only if they react weakly to self elements (and not if they do not react at all). Moreover, strong autoreactivity in the thymus can lead to the selection and differentiation of lineages specific to self elements, including regulatory T cells (cells that dampen the activation of the immune system and play a key role in the prevention of autoimmune diseases) (Wing and Sakaguchi 2010). In addition, many effector immune responses that occur routinely at the periphery during the lifetime of an organism target endogenous ("self") elements, as illustrated by the phagocytosis of dead cells, the clearance of cellular debris, many immune-mediated repair mechanisms, and the downregulating action of regulatory T cells, among many other phenomena. In other words, the claim that the immune system does not respond to self components is not true. There exists in fact a continuum from autoreactivity (interactions between immune receptors and endogenous motifs) to autoimmunity (the triggering of an effector response targeting endogenous motifs) and to autoimmune diseases (only the latter situation is pathological; it consists in the destruction of endogenous components, in a sustained manner and on a large scale – a given organ or even the whole organism in the case of systemic autoimmune diseases such as lupus) (Pradeu 2012).

Second, many genetically foreign entities are not eliminated by the immune system and are instead actively tolerated via regulatory immune responses. This includes examples such as foeto-maternal tolerance and various forms of chimerism, but also, and most crucially, immunological tolerance to a large number of bacteria, archaea, viruses, and fungi at all of the body's interfaces, including the gut, skin, lungs, sexual organs, and so on (Hooper and Gordon 2001; Chu and Mazmanian 2013; Virgin 2014; Chen et al. 2018). These resident entities, which are found in almost all living things in nature, are often referred to as the "microbiota." Contrary to what was thought for many years, these microbes are not "invisible" to the host immune system. The immune system interacts and develops an active dialogue with them, which leads to the triggering of a complex balance of effector and immunoregulatory mechanisms. In many cases, the host immune system facilitates the establishment and stability of certain components of the microbiota (Round and Mazmanian 2009; Donaldson et al. 2018). Importantly, microbial molecular patterns that were conceived for a long time as pathogenic signatures can also mediate tolerogenic immune responses (Sansonetti and Medzhitov 2009). All this confirms the centrality of the phenomenon of immunological tolerance, especially to the microbiota, in today's immunology (Pradeu and Carosella 2006a; Pradeu 2012).

Immunological interactions between host and microbes enable, in general, a peaceful coexistence between these two partners. It is estimated that a human host, for example, is made up of as many bacterial cells as its genetically self cells (Sender et al. 2016). Yet the most central point is not so much the number of these resident microbes as what they do in the organism. The involvement of bacteria in digestion has been known for decades, but recent research has shown that resident microbes play a major and sometimes even indispensable role in host activities as diverse as development, metabolism, defense, tissue repair (McFall-Ngai 2002; Xu and Gordon 2003; McFall-Ngai et al. 2013), as well as many other processes, including behavior (Sharon et al. 2016; Vuong et al. 2017; Schretter et al. 2018a). The fitness benefit for microbes consists generally in particularly favorable niches, where they find nutrients and protection from competitors, among other things (Donaldson et al. 2016). Microbes can also manipulate their host in many different ways (Sansonetti and Di Santo 2007). Finally, it is crucial to understand that the interaction between host and resident microbes is the product of a complex equilibrium, in which the nature of the ecological relationship can change through time depending on the circumstances and switch from mutualism to commensalism to parasitism (microbes that switch from the status of symbionts to pathogens have sometimes been called "pathobionts" (Chow et al. 2011)).

Recent work on immunological tolerance and the intimate dialogue between host and microbes across the living world invalidates the claim that the immune system eliminates foreign (nonself) entities. In the last two decades, symbiosis in general and symbiotic interactions with microbes more specifically have been recognized as ubiquitous and essential phenomena in nature. The specific contribution of immunology to this literature is to ask how symbiotic entities can be tolerated by the immune system and how they interact with this system in several central physiological processes (see Box 3.1).

Box 3.1 The Importance of Research on Symbiosis in Recent Biology and Philosophy of Biology and the Role of Immunology in That Research

Although research on symbiosis has a long history, the last two decades have seen a burst of investigations on this topic, at the crossroads of different domains, largely as a consequence of new technological advances (McFall-Ngai et al. 2013), particularly high-throughput sequencing and metagenomics. A key discovery was that virtually all living things harbor myriad microorganisms, many of which are beneficial to the host for activities as diverse as nutrition, development, and metabolism (Xu and Gordon 2003).

Recent work on symbiosis has also been of central interest to philosophers and conceptually interested biologists. Among the central issues, one can emphasize the degree of individuality displayed by host–microbe associations (Gilbert 2002; Pradeu and Carosella 2006b; Gilbert et al. 2012), the marriage of ecology and developmental biology (Gilbert 2002), an enrichment of the major transitions in evolution framework (Kiers and West 2015), the impact of symbiotic interactions on traditional views about evolutionary processes and the "tree of life" (Bouchard 2010) as well as debates about the concept of "holobiont" (Zilber-Rosenberg and Rosenberg 2008), and the long-recognized challenge of the transition from correlation to causality in microbiome studies (for example, if the microbiome composition in patients with inflammatory bowel disease differs from that of healthy individuals, is this difference a cause or a consequence of the disease?).

Not only has immunology been one of the most active fields in the development of this recent research on symbiosis (e.g., Hooper and Gordon 2001), but it also offers a specific perspective on this topic, namely the question of how so many genetically foreign entities can be tolerated rather than eliminated by the host immune system (Xu and Gordon 2003; Pradeu and Carosella 2006b; Round and Mazmanian 2009). In most cases, the symbiotic dialogue between the host and microbes is mediated by immune interactions (Round and Mazmanian 2009; Belkaid and Harrison 2017), which confirms the central role of the immune system as an interface with the environment as well as between components of the organism. One striking result is that microbes, long seen as what the immune system must reject, can often participate in the immunological defense of the host against pathogenic entities (Pamer 2016), thus creating a form of "co-immunity" (Chiu et al. 2017). This phenomenon of microbiota-mediated protection often involves components of the virome (all the viruses living in/on a host), which increasingly appears as a major future frontier in biomedical research (Virgin 2014).

The upshot is that the self–nonself theory is inadequate or at least incomplete, because many self components trigger immune responses and many nonself components are actively tolerated by the immune system. Therefore, today's immunology calls for novel and richer theoretical frameworks (Janeway 1989; Grossman and Paul 1992; Matzinger 2002; Pradeu et al. 2013), as well as renewed reflections about immunological individuality. The next section shows why, despite the critique of the self–nonself theory presented here, the claim that the immune system is pivotal for the definition of biological individuality remains entirely valid.

3.3 Immunology's Contribution to the Definition of Biological Individuality

At the most general level, the issue of biological individuality consists in asking what makes up a countable, relatively well-delineated, and cohesive entity in the living world (Sober 1991; Hull 1992). (Being countable and well-delineated concerns what can be called external unity, whereas cohesion has to do with internal unity.) Yet the meaning of countability, delineation, and cohesion remains unclear and is therefore a matter of debate among philosophers of biology and biologists. Accounts of these three dimensions have generally been based either on intuition or on evolutionary theory (Hull 1992; Godfrey-Smith 2009), leaving aside, in most cases, other biological fields, including physiological ones. In all these accounts the idea of "cohesion" seems particularly diverse and equivocal: a range of concepts have been suggested to elucidate this idea, including "functional integration" (the fact that parts are interconnected and interdependent) (Sober 1991), "near-complete decomposability" (the fact that interactions between parts of the individual are stronger than interactions between parts of the individual and the environment) (Simon 1969; Wimsatt 1972), and "cooperation and absence of conflict" (Queller and Strassmann 2009), but these concepts only partly overlap, and they are not always easy to define and to apply.

There is nowadays a quasi-consensus that biological individuality is question-dependent and practice-dependent, can be realized at different levels, comes in degrees, and should not be based on intuitions or on an anthropocentric approach (Pradeu 2016a; Lidgard and Nyhart 2017). I agree with this consensus. The specific question raised here is in what sense and to what extent the domain of immunology contributes to addressing the problem of biological individuality, essentially from a physiological point of view. (Section 5 discusses how to combine this approach with other approaches to biological individuality, including evolutionary ones.)

The main claim of this section is that immunology makes an important contribution to the definition of biological individuality insofar as it sheds light on all three dimensions discussed above, namely countability, delineation, and cohesion. The three main activities of the immune system participating in the individuation of biological entities can be called "filtering over entry," "filtering over presence," and "promotion of cooperation" (by "filter" I mean allowing or, on the contrary, restraining the entry or presence of something, as explained below). As Figure 3.1 indicates, these activities map onto countability, delineation, and cohesion.

"Filtering over entry" refers to the fact that the immune system constantly patrols the interfaces of the organism with the environment (skin, gut, and so on) and determines which exogenous elements can enter the organism and which can't. It thus plays a decisive role in the delineation of the organism's boundaries and the possibility of counting it as one single entity.

But the immune system does much more than control entries into the organism at interfaces. In a process that can be called "filtering over presence," the immune system constantly monitors the motifs expressed by the cells present in all tissues and body compartments as well as their intracellular content and

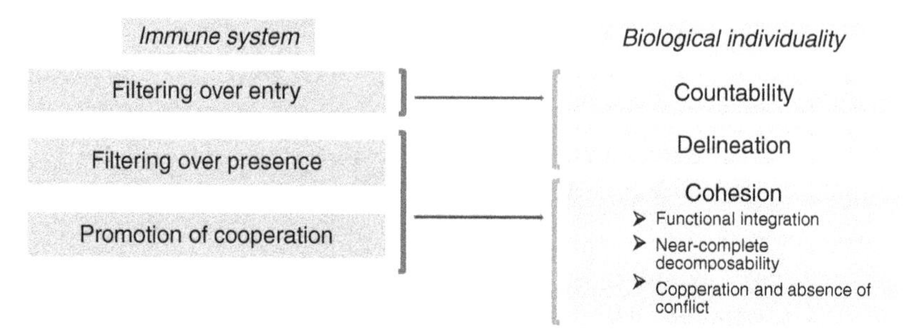

Figure 3.1 The three main activities by which the immune system participates in the individuation of biological entities and how they map onto the traditional conception of a biological individual. Filtering over entry sheds light on the idea of the individual as a countable and relatively well-delineated entity, while filtering over presence and promotion of cooperation shed light on the idea of the individual as a cohesive entity.

determines which elements are tolerated, and therefore can remain part of that living thing, and which elements are rejected, and therefore cannot remain part of that living thing. But what exactly does the immune system detect in this process of internal filtering? Although it has long been said that the immune system controls the identity of the elements with which it interacts (that is, their conformity with the self), my own view is that the immune system eliminates elements that change too abruptly, while tolerating elements that change slowly (Pradeu et al. 2013). Regardless of the criterion one adopts, though, it remains that the immune system constitutes such a filter over internal elements in addition to filtering the entry of external elements at bodily interfaces.

Lastly, the immune system plays an important role in the *promotion of cooperation* between the components of the organism. It does so in two different ways. First, the immune system can ensure long-distance communication between remote components of the organism. In plants this is done through the vasculature and thanks to metabolites (Shah and Zeier 2013). In metazoans this is made possible by different processes, most prominently the unique mobility of immune cells. For example, immune cells can inform distant organs about the presence of a pathogen in the organism and also contribute to tissue remodeling in remote sites (Eom and Parichy 2017). As we will see in Section 5, other systems like the endocrine and the nervous systems are also capable of long-distance communication in metazoans, but the immune system is unique in its capacity to send cells to every compartment of the organism. Second, the immune system can eliminate noncooperative elements ("cheaters"), for instance, cancer cells (see next section) (Michod 1999; Frank 2007), via the detection of abnormal molecular patterns, cell stress, aberrant cellular proliferation, and/or damage caused to the organism (Muraille 2013).

These last two activities (filtering over presence and promotion of cooperation) contribute to the cohesion of the biological individual (Figure 3.1). They strengthen functional integration (interconnection and interdependence), internal interactions (and therefore near-complete decomposability), as well as cooperation and absence of conflict between components.

Two important additional remarks are in order. First, the immune system is truly systemic, which means that, contrary to many bodily systems (e.g., digestive or respiratory), the immune system exerts its influence throughout the body. Therefore, filtering over entry, filtering over presence, and promotion of cooperation occur continuously and everywhere in the organism. (Many organs, either located at the interface with the environment or more internal, have been called "immunoprivileged" – for instance, the central nervous system and the eye, among many others. Yet recent research has shown that such compartments are not without immune influences (Mellor and Munn 2008; Louveau et al. 2015a). They are instead places where immunological processes are regulated differently than in the rest of the organism.)

Second, one crucial characteristic of the immunological approach to biological individuality is its broad range of application. It applies, in fact, to the whole living world, since immune systems are found in virtually all living things, including prokaryotes, plants, invertebrates, and vertebrates (see Section 2).

Thus, the immune system plays a central role in the delineation of the boundaries of a living thing and the determination of the cohesion between its constituents (Pradeu 2012). But is this account really different from the self–nonself framework examined above? It is actually essential to understand how the two accounts differ. For Burnet, the self–nonself framework is scientifically testable and useful if and only if self and nonself are defined in terms of *origins*: the self is what originates from the organism once it has acquired the capacity to recognize its own constituents, while the nonself is everything that does not originate from the organism. In stark contrast, the discrimination proposed above between what is part of the living thing and what isn't is not based on a question of origin, since in this view many genetically foreign entities can be accepted by the immune system while many endogenous entities are routinely destroyed. For example, in the account developed here, microbiota components that are immunologically tolerated belong to the organism, while they are considered nonself and therefore not part of the organism in the self–nonself theory. In other words, I propose that the immune system contributes to the continuously re-delineated distinction between the "inside" and the "outside" but this distinction is not to be confused with the opposition between the "endogenous" (what comes from the inside) and the "exogenous" (what comes from the outside). The resulting living thing does not coincide with the traditional self and cannot be accounted for by Burnet's self–nonself theory.

A related idea is that the process of dual filtering presented here is dynamic and never-ending. There is, therefore, a constant re-delineation, through the action of the immune system, of the constituents and boundaries of a living thing. This sheds a crucial light on the *diachronic* identity of biological individuals (Wiggins 2016). An entity that is part of a living thing at a given moment (for example, an immunologically tolerated virus) can cease to be part of that living thing later in time (for example, if this virus is eventually eliminated by the immune system). The resulting individuality is relative and changes continuously, but the criterion used to establish what is part and what is not part of a living thing remains constant and offers a precise delineation.

3.4 The Role of the Immune System in Turning a Set of Heterogeneous Constituents into an Integrated Individual

Immunology helps us better understand what makes a living thing a highly integrated individual despite the diversity of its constituents. (Like many other aspects of biological individuality, integration – which we defined as one way of conceiving cohesion – is a matter of degree.) According to the immunological account proposed here, every living thing is a chimera, a heterogeneous and mixed entity composed of genetically diverse components (Pradeu 2010; Dupré 2010). On the one hand, it is a composite entity, comprising biotic elements that originate both from the organism and from outside the organism (bacteria, viruses, fungi, and so on). On the other hand, it is a very special composite entity, exhibiting a high degree of integration, with well-delineated boundaries, tight interactions, and strong cooperation between its components. Because of its three key activities of filtering over entry, filtering over presence, and promotion of cooperation, the immune system is essential in determining the constituents and boundaries of the living thing and therefore in turning a set of heterogeneous components into an integrated individual. In other words, the immune system exerts a major "e pluribus unum" activity: it constantly turns a plurality of elements that are diverse and of various origins into a cohesive unit. From this point of view, the immune system, along with other (and closely connected) "policing" mechanisms (such as apoptosis) and systems pertaining to adhesion and intercellular communication, participates in the "glue" of life, the sticking together of diverse elements that eventually constitute a physiological unit. (For a similar claim centered on the specific example of sponges, see (Müller 2003).)

This reasoning allows us to propose an immunological definition of the organism, understood as a physiological individual (i.e., a cohesive whole, functioning collectively as a regulated unit that persists through time). According to this definition, an organism is a continuously changing, functionally integrated whole made up of heterogeneous constituents (including many microbes) that are locally interconnected by strong biochemical interactions and controlled at a systemic level by immunological interactions (Pradeu 2010). Everything that is actively interacting locally and tolerated by the immune system is part of the physiological individual.

This means that immunological processes can help us identify and delineate physiological individuals not via intuition and common sense (as has often been a concern with physiological definitions) but on the basis of a clearly formulated and scientifically grounded criterion (following the argument of (Hull 1992)). The boundaries of the immunologically delineated living thing may be a matter of degree and vary depending on the context, but the criterion itself is precise and clear. For example, there are uncertainties and debates over what counts as a biological individual in the case of a colonial entity such as the ascidian *Botryllus schlosseri* (Figure 3.2): each cell, each zoid, the whole colony, or perhaps all of them together (Buss 1999)? If we adopt the immunological perspective suggested here, then the whole colony should be recognized as the physiological individual because it is at the level of the colony that immunological processes occur such as histocompatibility-based fusion/rejection responses when two colonies meet (De Tomaso et al. 2005). Similarly, there are many discussions about whether or not the microbiota is part of the individual it inhabits (e.g., (Gilbert et al. 2012)). Our immunological account tells us that all (and only) the microbes that are immunologically tolerated are part of the physiological individual.

Crucially, organismality here is a product, not a given. Indeed, my claim is *not* that the components of an already identified organism are tolerated by its immune system. Instead, I propose that it is the identification of a local concentration of biochemical and immunological processes taken together that tells us how to pick out organisms (understood as physiological individuals) in nature on the basis of the capacity of the immune system to determine what the constituents and boundaries of a living thing are. We start with an unindividuated living mass (*Botryllus schlosseri*, for example), but the identification of immunological processes allows us to single out and delineate quite precisely a physiological individual.

Biological individuation, as we saw, can be realized at different levels. An interesting aspect of immunological individuation is that it also can occur at different levels. Multicellular organisms exhibit what can be called a *multilevel*

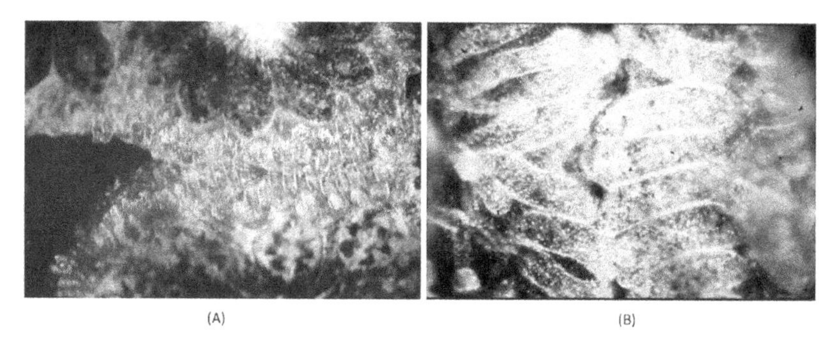

Figure 3.2 Rejection between two colonies of *Botryllus schlosseri*. When two colonies of *Botryllus schlosseri* meet, they can reject (panel A) or fuse (not shown). This occurs at the colony level. The brown zones show the starting point of rejection. Panel B shows rejection at the more precise level of ampullae. (Photographs courtesy of Tony De Tomaso, UCSB)

immunity: key immunological processes occur at the level of each cell (for example, any cell of the organism infected by a virus will respond by triggering a diversity of protective mechanisms, including the production of interferons and other cytokines – a phenomenon often referred to as cell intrinsic immunity – which can alter neighboring cells about the presence of the virus) (Goubau et al. 2013) and at the level of each tissue. Most of the time, one can distinguish degrees of immunological responses: in most multicellular organisms, for instance, cell and tissue immunological responses are strongly controlled and coordinated at the systemic level, so the systemic level seems to be the level at which the highest degree of immunological integration is realized. Interestingly, in some cases, the highest degree of immunological integration is realized at the level of a colony rather than intuitively defined individuals. For example, some data in a number of social insects such as termites and honey bees suggest that key immunological processes occur at the colony level, which has led to the concept of "social immunity" (Cremer et al. 2007; Jones et al. 2018) and has sometimes been used to support the superorganism hypothesis.

The conclusion of this discussion is that recent immunological research has seen extensive revisions of its core concepts (particularly "self," "tolerance," and "microbe"), which in turn have led to a significant reassessment of our traditional understanding of biological individuality insofar as a living thing can be seen as an immunologically unified chimera (Gill et al. 2006; Eberl 2010; Bosch and McFall-Ngai 2011; Pradeu 2012). But do these changes have any practical consequences? The short answer is that they have many practical consequences, particularly from a therapeutic point of view. Let's take two examples of this – autoimmune diseases and ecological approaches to the microbiome.

If the perspective presented here is correct, then autoimmunity is a physiological process, which plays indispensable roles in the organism. Autoimmune diseases must be perceived not as the result of the sudden appearance of an undesirable autoimmune reaction (in the tradition of Ehrlich's "horror autotoxicus") and elimination of immune cells responding to the self, but as the consequence of a perturbation in the (cell-mediated, tissue-mediated, or system-mediated) regulation of immunological processes that include, as a normal component, responses to endogenous elements (Wing and Sakaguchi 2010). This means that, contrary to what has been done over the last decades, the first question of researchers interested in understanding autoimmune diseases should perhaps be not so much "what triggers an immune response against the self?" (e.g., why do self-reactive effector T cells develop in this organism?) as "which aspect of the regulatory machinery that normally keeps autoimmune responses in check has been disturbed, and why?" (e.g., why do regulatory T cells no longer downregulate the activity of potentially harmful autoreactive T cells?). The two strategies differ. The first acts at the level of effector cells and consists in preventing the development of autoreactive cells. The second acts at the level of the cells that regulate the activity of effector cells; acknowledging the physiological nature of autoimmunity, it consists in preventing the switch from physiological autoimmunity to pathological autoimmunity. One can speculate that many factors could play a role in this disturbance of regulatory mechanisms (especially regulatory T cells), including excessive hygiene in industrialized societies (Bach 2002), modification of the microbiota, and disorganization of the extracellular matrix in local tissues.

The second example of a therapeutic consequence of the renewed conception of immunity and individuality in recent immunology concerns the microbiota. Recognizing that most living things harbor myriad microbes, most of which do not harm their hosts, a key clinical goal becomes to manage these microbial communities instead of trying to kill all microbes. The development of such ecological approaches in medicine is a difficult task because the microbiota is in fact made of several highly complex local ecosystems in which any perturbation is likely to have unpredictable consequences. That being said, recent success with faecal transplantation in people infected with *Clostridium difficile* (van Nood et al. 2013), despite major potential limits in other contexts (Pamer 2014), suggests that such approaches are of increasing interest to clinicians. Another approach in the same vein is to determine to what extent, in a context of antibiotic resistance, it will be possible to manipulate components of the microbiota to protect a host against pathogens (Pamer 2016).

3.5 Combining Different Approaches to Biological Individuality

Immunology is clearly not alone in shedding light on biological individuality, an issue to which many biological fields can contribute (Clarke 2011; Pradeu 2016a; Lidgard and Nyhart 2017). First, the contribution of immunology, as we saw, concerns primarily the biological individual understood from a *physiological* point of view, and moreover it is not the only field contributing to that question (developmental biology, neuroscience, and many others also contribute). Second, beyond physiological fields, other biological domains, most prominently evolutionary biology, can shed light on the problem of biological individuality. In the last four decades or so, especially after the founding work of David Hull, most of the debates about biological individuality among philosophers of biology and biologists have focused on evolutionary individuality. This includes discussions over units of selection, the replicator/ interactor distinction (Hull 1980), Darwinian individuals (Godfrey-Smith 2009), the possibility of defining species as individuals (Hull 1978; Haber 2016), the question of how groups of entities can aggregate and form new individuals (Buss 1987; Maynard Smith and Szathmáry 1995; Michod 1999), as well as multilevel selection and the constitution of an organism from an evolutionary viewpoint (Okasha 2006; Queller and Strassmann 2009). Other fields also have much to say about the concept of a biological individual, including, for instance, ecology (Huneman 2014).

Recently there have been many calls in favor of pluralistic and practice-oriented approaches to biological individuality (Kovaka 2015). As useful as these calls have been in undermining the idea that accounts of biological individuality based solely on a single scientific field (typically evolution, but the same would apply to any other "monistic" approach) and on a purely theoretical viewpoint would be sufficient, it is now important to take a step further. The existence of a plurality of approaches to biological individuality, many of which are rooted in certain scientific practices, is not disputable. The major challenge at present is to determine whether it could be useful to *combine* different approaches to biological individuality and, if so, how (Pradeu 2016b; Lidgard and Nyhart 2017). Some entities in the living world display a high degree of physiological and/or metabolic individuality without displaying a high degree of evolutionary individuality, or the other way around (Dupré and O'Malley 2009; Pradeu 2010). Other entities, in contrast, may express a high degree of individuality on both grounds. Perhaps determining which entities express a high degree of individuality along several criteria tells us something important about their roles in the living world. An equally important issue is to establish to what extent different criteria of biological individuality are compatible or mutually exclusive. Indeed, even if scientific practices of individuation vary from one scientific field to another, there often is a significant overlap between the questions and criteria used by different biologists when they talk about biological individuals (Guay and Pradeu 2016). This constitutes a promising research program for philosophers of biology interested in making comparisons across several scientific fields.

In summary, the immune system plays a major role across species in determining the boundaries, constitution, and cohesion of the biological individual. Far from the original concepts of self and nonself, today's immunology tells us that a living thing can be seen as an immunologically integrated chimera. Although certainly not unique, the contribution of immunology to the long-standing debate over biological individuality is important, so much so that philosophers and biologists interested in that debate cannot neglect this contribution. What immunology has to say about individuality has long-reaching consequences for our understanding of what a living thing is, including what we are as humans, and it also impacts health issues of central importance, from autoimmune and inflammatory diseases to antibiotic resistance and transplantation.

4 Cancer as a Deunification of the Individual

> Cancer kills millions of people every year and is one of humanity's greatest health challenges. By stimulating the inherent ability of our immune system to attack tumor cells this year's Nobel Laureates have established an entirely new principle for cancer therapy. Allison and Honjo showed how different strategies for inhibiting the brakes on the immune system can be used in the treatment of cancer. The seminal discoveries by the two Laureates constitute a landmark in our fight against cancer. (Excerpt of attribution of the 2018 Nobel Prize in Physiology or Medicine).

"Why don't we get more cancer?" asks Mina Bissell, a prominent specialist in cancer (Bissell and Hines 2011). Certainly, most of us think that we do see enough cancer around us. After all, approximately 90 million people had cancer in 2015 (Vos et al. 2016) and cancer kills around 8.8 million people each year (Wang et al. 2016). The 2014 World Cancer Report of the World Health Organization estimated that there were about 14 million new cases of cancer each year and that the financial costs of cancer were above US $1.16 trillion per year. In this context, asking why we don't get more cancer may seem surprising if not shocking.

Yet Bissell's question is entirely legitimate. We probably all have occult tumors (Bissell and Hines 2011) – what Folkman and Kalluri called "cancer without disease" (Folkman and Kalluri 2004). Autopsies of people dead due to reasons unrelated to cancer reveal the high frequency of tumors, which can be large but do not spread and do not seem to threaten the host. Prostate tumors can be found in 30 to 70 percent of men over 60 years old, breast tumors in 7 to

39 percent of 40- to 70-year-old women, and thyroid tumors in 36 to 100 percent of 50- to 70-year-old adults (Welch and Black 2010).

So, if tumors are so frequent, why do only a limited number of them turn into actual cancers? The reason we don't get more cancer is that several systems of regulation exist in the organism, systems that prevent the formation of a cancerous tumor, its growth, and its invasion of distant tissues (metastases) (Klein et al. 2007). These regulation mechanisms are diverse and located at different molecular, cellular, and tissue levels (Hanahan and Weinberg 2011): they include DNA maintenance and repair (Hoeijmakers 2001), regulation of cell cycle (Sherr 1996), apoptotic signals sent to cancer cells by surrounding cells (Letai 2017), and inhibitors of angiogenesis (formation of new vessels) (Jain 2005), among many others.

The aim of the present section is to show that, among these regulation mechanisms, the immune system plays a critical role (Binnewies et al. 2018) and that, conversely, the immune system is certainly involved in every cancer. The crucial point here is that cancer is a disease of multicellularity – or, more precisely, cancer is a disease that results from a dysfunction of the mechanisms that normally insure the cohesion of the multicellular individual, the very mechanisms among which we have seen that the immune system is crucial (see Section 3). A striking result of research done in the last twenty years is that the immune system can both *prevent* and *support* cancer formation (de Visser et al. 2006; Binnewies et al. 2018). Understanding how the immune system, which we characterized in previous sections as essential for maintenance of the cohesion of the organism, can in some circumstances promote cancer constitutes a major challenge from a conceptual as well as a therapeutic point of view.

The outline of the section is as follows. First, I explain the progressive construction of the idea that the immune system can restrain cancer. Second, I explore how the immune system was subsequently described as capable of both restraining and promoting cancer. Third, I sum up current views about immune–cancer interactions and their clinical applications, especially immunotherapies. Finally, I examine the role of the immune system in the breakdown of biological individuality often described as typical of cancer.

4.1 How the Immune System Restrains Cancer: The Complex History of the Idea of Immunosurveillance

It is obviously difficult to make broad generalizations when discussing cancer. Most of the time, when we talk about cancer, we refer to a set of significantly heterogeneous diseases characterized by uncontrolled cell multiplication and the cells' potential to invade other parts of the organism. The heterogeneity of cancer is a well-established point (Melo et al. 2013; Bertolaso 2016; Plutynski 2018). First, cancers found in humans are quite diverse. Solid tumors (e.g., a carcinoma, a cancer that develops from epithelial cells, which constitutes the most frequent form of cancer) differ from liquid tumors (e.g., leukemia, a cancer that develops from bone marrow–derived cells). Cancers also differ from organ to organ: for example, a breast cancer can be quite unlike a skin cancer (this is called intertumor heterogeneity) (Melo et al. 2013). Cancers affecting a given organ, such as the breast for example, can have different characteristics (Zardavas et al. 2015). Moreover, a cancerous tumor can exhibit a high level of internal heterogeneity ("intratumoral heterogeneity") (Fisher et al. 2013). Last but not least, some tumors are benign, as they do not invade other body parts. Second, beyond humans, cancers exist in a great number of multicellular organisms, which further increases the diversity of cancer types and characteristics. If cancer is defined as a breakdown of multicellular cooperation that manifests itself by uncontrolled proliferation, inappropriate cell survival, resource monopolization, deregulated differentiation, and degradation of the environment, then it is found in many different groups, including starfish, hydra, insects, and plants (Aktipis et al. 2015).

Despite the diversity and heterogeneity of cancers, the immune system has been shown to be implicated in cancer progression in a large range of cancer types in various vertebrates (and also, if perhaps more speculatively given the limited number of studies, in invertebrates (Pastor-Pareja et al. 2008; Robert 2010)). Because of the publicity around cancer immunotherapies, most people are now aware that the immune system plays a role in cancer. Yet much remains to be done to understand the detailed mechanisms underlying this process. Study of the interactions between cancer and the immune system has undergone major transformations in the last two decades. To understand why the immune system can participate in the rupture of cohesion that characterizes cancer, it is essential to understand the nature of these recent transformations.

The history of the study of how the immune system influences cancer progression is itself illuminating. In the 1890s William Coley (Coley 1893) took advantage of the observation that spontaneous tumor regression could follow infection with a pathogen to develop a killed bacterial vaccine for cancer, a phenomenon in which immune components were thought to play a role (Cann et al. 2003). At the beginning of the twentieth century, Paul Ehrlich suggested that the immune system could recognize and eliminate malignant cells (Ehrlich 1909). In the mid-1950s, Lewis Thomas (at a symposium held in 1957, and published two years later (Thomas 1959)) and Macfarlane Burnet (Burnet 1957) argued that cellular immunity was, in the words of Thomas, "designed as a useful and effective mechanism for the early sensing and early elimination of neoplastic cells" (Thomas 1982, p. 330). Thomas was convinced that humans produce tumors all the time but keep them under control, thanks to the action of the immune system. Burnet later expanded this idea and coined the term immunological surveillance (sometimes called *immunosurveillance*) (Burnet 1970). According to immunological surveillance, the immune system is capable of detecting and eliminating the "altered self" (Houghton 1994), that is, "new antigens" (Burnet 1970, p. 7) (antigens that are different from those of the body, also called "neoantigens" (Schumacher and Schreiber 2015)). In the 1970s, the idea of immunological surveillance started to decline because several experiments suggested that immunodeficient mice did not have a higher susceptibility to spontaneous or chemically induced tumors (Stutman 1974). As a result it was almost entirely abandoned for several years. A clear indication of this is the fact that the highly influential review on the "hallmarks of cancer" by Hanahan and Weinberg published in 2000 (Hanahan and Weinberg 2000) ignores the role of the immune system in cancer (this was corrected in (Hanahan and Weinberg 2011)).

At the beginning of the 2000s, a series of experiments showed the impact of the immune system on cancer development (Shankaran et al. 2001; Dunn et al. 2002). The supposedly immunodeficient mice used in experiments from the 1970s onward to invalidate the idea of immunosurveillance in fact had an immune system (they had NK cells, $\gamma\delta$ T cells, and even some $\alpha\beta$ T cells) (Dunn et al. 2002). During this period, the involvement of both innate and adaptive immune components in cancer control was demonstrated (immunochemical or functional ablations of NKT, $\gamma\delta$ T cells, NK cells, $\alpha\beta$ T cells, IFN-γ, and interleukin 12 all lead to increased susceptibility to cancer).

4.2 The Dual Action of the Immune System in Both Restraining and Promoting Cancer: Immunoediting and Beyond

Although much emphasis has been put on the capacity of the immune system to restrain cancer, the immune system can also, perhaps paradoxically, promote cancer (de Visser et al. 2006). Here again, both the innate and adaptive components of the immune system can participate in such cancer-promoting processes. One important step in the realization that the immune system can favor cancer development was the switch from the concept of immunosurveillance to that of immunoediting (Dunn et al. 2002). Several researchers working on immunosurveillance noted that the action of the immune system could lead to the selection of more resistant tumor cells. Highly immunogenic tumor cells are eliminated by the immune system, but this process leaves behind tumor variants of reduced immunogenicity (or that have acquired other mechanisms to evade or suppress the immune response) and have a higher fitness in the immunocompetent host (a process similar to the selection of more resistant pathogens). Therefore, Robert Schreiber and colleagues proposed to use the term immunoediting instead of the traditional notion of immunosurveillance, reflecting the fact that the immune system not only monitors tumors but also shapes them, with both beneficial and detrimental consequences for the host (Dunn et al. 2002; Schreiber et al. 2011). Immunoediting is a more encompassing and more accurate concept, especially because it emphasizes the diachronic character of cancer development.

The concept of immunoediting is well-suited to describe the dual host-protecting and tumor-sculpting actions of the immune system. According to its proponents, immunoediting encompasses three different though largely overlapping processes, referred to as the "3 e's" (see Figure 4.1): "elimination" (which corresponds to the classic idea of immunological surveillance, that is, the destruction of tumor cells by the immune system), "equilibrium" (the immune system iteratively selects and/or promotes the generation of tumor cell variants with increasing capacities to survive immune destruction), and "escape" (the immunologically sculpted tumor expands beyond control in the host) (Dunn et al. 2002). In 2007, a landmark study by Schreiber's group confirmed the existence of the equilibrium phase (Koebel et al. 2007).

The processes by which the immune system can have a tumor-promoting effect go well beyond immunoediting and are actually quite diverse. In particular, tumor-associated macrophages (TAMs) can sometimes be tumor-suppressive, but in a majority of cases they favor tumor initiation, progression, and metastasis (Mantovani et al. 1992; Wynn et al. 2013). Indeed, macrophages can sustain the chronic inflammation that often plays a role in tumor initiation and promotion (Mantovani et al. 2008), skew adaptive immune responses, and facilitate cell growth, angiogenesis (Murdoch et al. 2008), matrix deposition, tissue remodeling (Mantovani et al. 2013; Afik et al. 2016), and metastasis (Qian and Pollard 2010). The exact effects of macrophages located in or around the tumor also depend on the temporal sequence of events: tumor-preventing macrophages can switch to a tumor-promoting role depending on the cues they receive from the tumor microenvironment (Wynn et al. 2013). So-called myeloid-derived suppressor cells (MDSCs) contribute to cancer and metastasis (Kumar et al. 2016), and tumor-associated neutrophils (Fridlender et al. 2009) as well as regulatory T cells (Tanaka and Sakaguchi 2017) also, in many circumstances, can promote cancer.

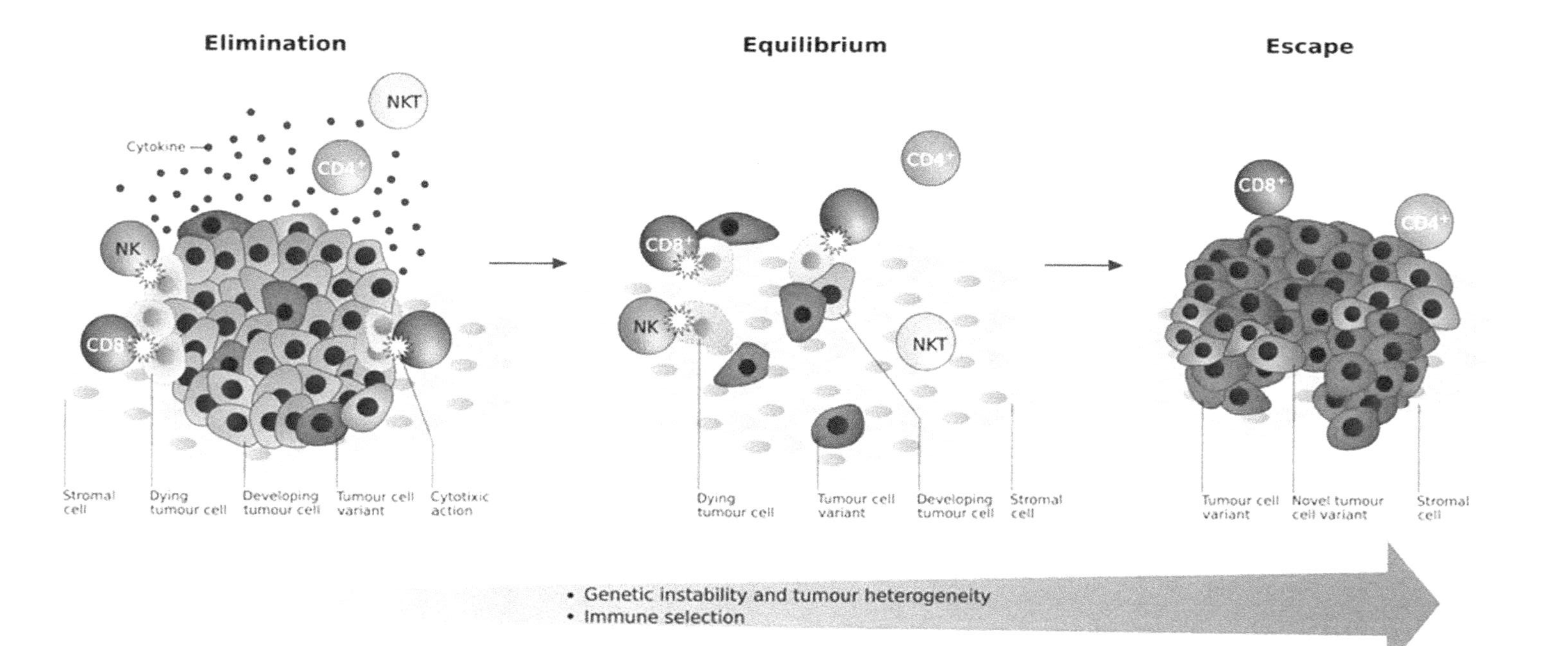

Figure 4.1 The three phases of immunoediting. According to the concept of immunoediting, three phases must be distinguished in the interactions between the immune system and the tumor: (a) *elimination*, which corresponds to the destruction of tumor cells by the immune system; (b) *equilibrium*, in which the immune system selects and/or promotes the generation of immunologically resistant tumor cell variants; and (c) *escape*, which corresponds to the expansion of the immunologically shaped tumor that is now beyond the control of the immune system. (Figure drawn by Wiebke Bretting, after (Dunn et al. 2002)).

4.3 Immune–Cancer Interactions: Current Views and Clinical Applications

Today, the evidence showing that the immune system plays a central role in cancer development (either preventing or promoting cancer) is overwhelming (Schreiber et al. 2011; Chen and Mellman 2017; Ribas and Wolchok 2018). Many cellular and molecular immune components are involved in this process of restriction or promotion of cancer. Additionally, these interventions of the immune system can occur at different levels in or around the tumor (including genes, cells, and the tumor microenvironment) and at every step of its progression, from initiation to neoplastic progression to metastasis. For example, mouse models showed that tumor-associated macrophages promote angiogenesis and tissue remodeling, thereby favoring tumor growth. Clinical studies show that extensive tumor-associated macrophage infiltration positively correlates with cancer metastasis and poor clinical prognosis (Afik et al. 2016)

Investigations about the role of the immune system in cancer progression have led to remarkable clinical applications. For example, several groups have shown how, in a diversity of cancers, infiltration by certain immune components had a better prognostic value than more traditional approaches (Galon et al. 2006). But most crucially, since the 2010s, the knowledge accumulated over decades about immune-mediated control of cancerous tumors has turned into specific clinical applications called "cancer immunotherapies," which many have described, rightly or wrongly, as "revolutionary" (Kelly 2018). After extremely encouraging results obtained in melanoma and a few other cancer types at the beginning of the 2010s (Hodi et al. 2010; Wolchok et al. 2013), evidence in favor of the success of immunotherapies (increasingly, in fact, a combination of immunotherapies) in several cancers has accumulated (Ribas and Wolchok 2018). In a number of situations, the results have been unprecedented and sometimes even spectacular, especially in cases of previously incurable cancers, raising much enthusiasm. Immunotherapies are diverse but recently immune checkpoint inhibitors have been particularly explored (Leach et al. 1996; Ribas and Wolchok 2018).

The most significant recognition of the work done in this area is undoubtedly the 2018 Nobel Prize in Physiology/Medicine awarded to James P. Allison and Tasuku Honjo "for their discovery of cancer therapy by inhibition of negative immune regulation," which is centered on the blockade of immune checkpoints. This has definitely convinced researchers working in all areas of cancer investigation and the lay public that it is indispensable to pay attention to the role of the immune system in cancer. It comes as no surprise that many newspapers have discussed these medical advances, if perhaps sometimes hyperbolically (e.g., (Vonderheide 2018)).

Are immunotherapies really revolutionary, and what do they tell us about immune–cancer interactions? From a strictly medical point of view, this enthusiasm is justified, although we should keep in mind that some important limits exist. The first limit is the low percentage of responders: less than 15 percent on average, though it depends on tumor type and on the category of immunotherapy (Haslam and Prasad 2019). (Future research likely will significantly extend the proportion of responders.) A second limit is the existence of sometimes significant adverse effects (immunotherapies, in particular, increase the level of inflammation and autoimmune responses, which can lead to colitis, hepatitis, etc.) (Postow et al. 2018). A third limit is the currently exorbitant cost of some treatments.

From a conceptual viewpoint, therapies based on immune checkpoint inhibition constitute indeed a radical change in perspective (Lesokhin et al. 2015; Sanmamed and Chen 2018). At least two important features of immune checkpoint therapies are worth emphasizing. First, *the target of the treatment is the immune system*, not the tumor itself (as was the intention with traditional treatments such as surgery, chemotherapy, and radiotherapy – although, in fact, some of them are now known to act at least in part via stimulation of the immune system (Galluzzi et al. 2012)). Second, the objective is to *break the state of immune tolerance* that has been established between the tumor and the immune system in the local tissue (Lesokhin et al. 2015; Ribas and Wolchok 2018) – partly as a consequence of the chronic expression of cancer antigens (Pauken and Wherry 2015). More precisely, the aim with immune checkpoint blockers is to downregulate inhibitory signals in tumor–immune interactions. This constitutes a move from enhancement of the immune system to "normalization" of the immune system: in immune checkpoint inhibitor-based immunotherapies, especially with anti-PD-L1, the aim is not, strictly speaking, to boost the immune system beyond its normal rate of activation but to restore a local context in which the immune system will be able to act as it normally does (Sanmamed and Chen 2018).

Together, basic studies about the role of the immune system in cancer and clinical studies in the domain of immunotherapies also have contributed to an important change in perspective about what cancer is and how it develops (Prendergast 2012). It is increasingly recognized that tumor-centric views of cancer (with genetic mutations seen as the main cause of cancer) are insufficient: to understand (and cure) cancer, it is essential to consider not only the tumor itself but also the tumor environment ((Bissell and Hines 2011; Maman and Witz 2018); for conceptually and philosophically-oriented analyses, see (Bertolaso 2016; Laplane 2016; Plutynski 2018)). The tumor environment includes the tissue context located at the vicinity of the tumor (sometimes called the tumor microenvironment), but also elements located quite remotely from the tumor in the organism (such as some immune-associated organs and the microbiota, which recently has been proven to influence cancer progression and therapies) (Zitvogel et al. 2018; Binnewies et al. 2018; Laplane et al. 2018). Even authors who initially focused on intrinsic molecular aspects of cancer development have later emphasized the importance of the tumor microenvironment (Hanahan and Weinberg 2011). Targeting the tumor microenvironment also offers enriched therapeutic strategies (e.g., (Joyce 2005)). There is a growing consensus that the immune system plays a crucial role in the tumor microenvironment (Bissell and Radisky 2001; Binnewies et al. 2018; Maman and Witz 2018). In fact, given the centrality of immune components in the organization of, and control over, the local tissue, it seems reasonable to say that every cancer involves the immune system, which necessarily intervenes, at one point or another, in the shaping of the local context that enables the tumor to emerge, grow, and perhaps spread.

4.4 Role of the Immune System in the Breakdown of Biological Individuality That Characterizes Cancer

With all this discussion over the tumor-restricting and tumor-promoting roles of the immune system in mind, we can now return to the question with which we started. What exactly is the role of the immune system in the prevention of the breakdown of individuality that characterizes cancer, and how can the immune system be involved, conversely, in the decohesion of the biological individual?

The idea that cancer constitutes a breakdown of biological individuality is widespread in the scientific and philosophical literature. Biologist Leo Buss was instrumental in showing that biological individuality in multicellular organisms must be understood as an outcome of evolution, by which, on several occasions in life's history, some cells aggregated and cooperated, and in which emerged some control mechanisms over cells that would proliferate at the expense of the whole organism (Buss 1987). Buss takes cancer as an example of a decohesion of the biological individual, in which cancer cells are "re-individualized" in a way that becomes harmful to the multicellular organism. This idea has subsequently been explored by several biologists and philosophers of biology (Frank 2007; Germain 2012; Plutynski 2018), often inspired by the study of clonal evolution at the cell level in cancer (Nowell 1976; Greaves and Maley 2012). For example, Godfrey-Smith labels as "de-darwinization" the process by which a higher-level individual prevents proliferation of lower-level individuals (Godfrey-Smith 2009, pp. 100–103). From that point of view, cancer cells appear as a result of a "re-darwinization" at the cell level.

However, the mechanistic details by which the multicellular organism exerts control over cancer cells have remained vague. Michod (Michod 1999, p. 119) cites programmed cell death and the immune system as the two main "policing mechanisms" in the multicellular organism, but he does not give any detailed explanation about how they work. An important lesson of what has been said in this section is that immunological surveillance constitutes a convincing example of a mechanistically precise process of maintenance of cohesiveness in the organism (Prendergast 2012; Pradeu 2013). Thus, it offers an important contribution to this long-standing debate concerning de-darwinization in cancer. The details of how immune-mediated control works are well documented: in the elimination phase of immunoediting, myriad immune cells and molecules (macrophages, dendritic cells, NK cells, $\gamma\delta$ T cells, CD4 and CD8 T cells, IFN-γ, among many others) contribute to the destruction of the tumor (Dunn et al. 2002). In addition, immune-mediated control contributes to coordinate other control instruments, such as apoptosis and angiogenesis.

This, however, strengthens rather than solves the paradox: if immune-mediated restriction is one of the main mechanisms ensuring the cohesion of the organism, we need an explanation for why in some circumstances the immune system *favors* tumor development (through the escape phase of immunoediting and/or through repair mechanisms that contribute to creating a favorable tissue environment for the tumor). To better understand this phenomenon, I propose here an extended view of immune-mediated cohesion and decohesion.

The traditional view about immune-mediated cohesion, as represented in classically defined immunological surveillance, is that the immune system can eliminate abnormal cells such as cancer cells (Figure 4.2).

This view, however, is too narrow, because it neglects the diversity of activities in which the immune system is involved, which include not only defense but also development, tissue repair, clearance of debris, and maintenance of tissue homeostasis, among others (see Section 2). In cancer, many of these immunological activities are found. This leads to a much richer view of immune-mediated cohesion and decohesion in cancer (Figure 4.3). In this view, the immune system plays a major role in regulation of the organization of the local tissue, and, together, the immune system and the tissue realize different activities, including the elimination of abnormal cells, but also the *containment* of abnormal cells (in that case, cells are not destroyed, they are simply kept under control, which limits the damage they can do and/or their capacity to

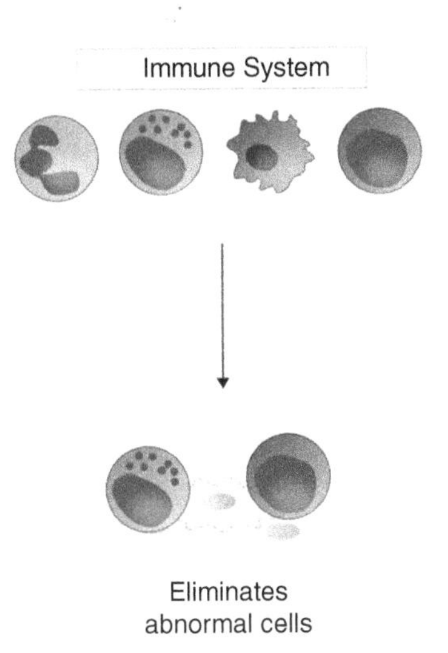

Figure 4.2 The traditional view of immune-mediated cohesion, as proposed by the immunological surveillance hypothesis. According to this view, the immune system directly eliminates abnormal cells, such as cancer cells (in red) in the tumor. (Figure drawn by Wiebke Bretting).

spread), the *maintenance* of chronic elements present in the local environment (most of the time, these chronic elements are normal self components of the organism, but chronically present tumor constituents can also lead to active maintenance of the tumor by the immune system, which progressively sees these elements as normal) (Pradeu et al. 2013; Pauken and Wherry 2015), and the *repair* of the local tissue (in physiological conditions, this repair is a necessary process insuring the integrity of the tissue, but in the context of cancer repair mechanisms can favor cancer progression).

The crucial point here is that decohesion as seen in cancer can concern all these different activities, not just elimination: the immune system, which in most cases prevents cancer progression by elimination, containment, maintenance, and repair, can in some circumstances promote cancer progression because of deregulated elimination, containment, maintenance, and/or repair. Moreover, all these activities must be understood diachronically: they do not all intervene at the same time, and the immune system can switch from one effect to the other (for example, it can initiate the destruction of abnormal components and later contribute to their maintenance because they have become chronically expressed in the tissue).

Does the immune system dysfunction when it promotes cancer progression via a deficit in elimination or containment and/or via maintenance processes and/or via repair processes? I suggest distinguishing two situations here. The

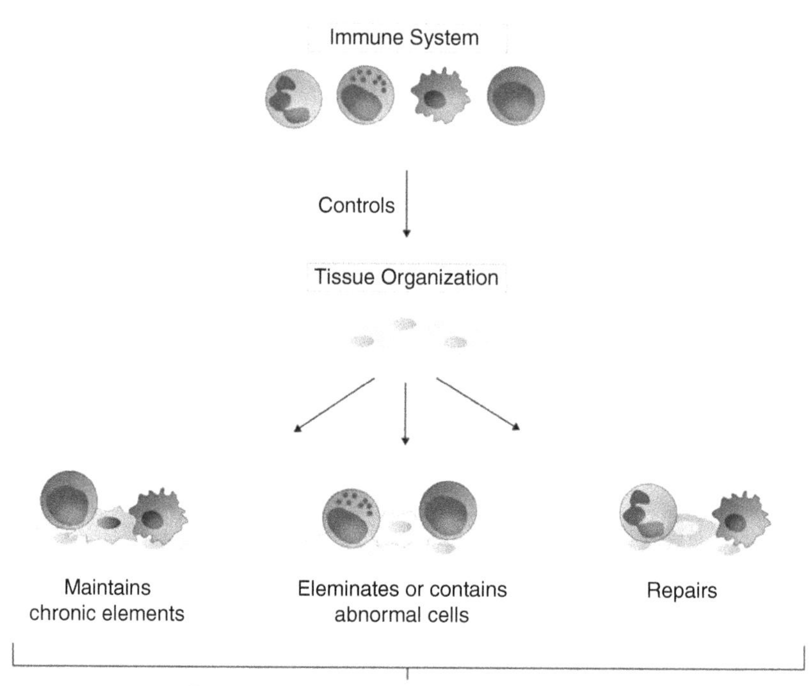

Figure 4.3 A richer view of immune-mediated cohesion and immune-mediated decohesion in cancer. In this view, the immune system controls tissue organization and, together, the immune system and the local tissue can exert a variety of cohesion-promoting activities, including the elimination of abnormal cells, but also the containment of abnormal cells, the maintenance of chronic elements, and tissue repair. All these activities (not just elimination), in pathological conditions, can promote decohesion of the organism. (Figure drawn by Wiebke Bretting).

first situation corresponds to a dysfunctional immune system. An organism, either structurally or provisionally (e.g., after a treatment with immunosuppressive drugs), can have a defective immune system (e.g., a deficit in effector T cells, or a disequilibrium in the respective numbers of its inflammatory and regulatory macrophages – or, more specifically, of its macrophages distributed along the inflammatory to "alternatively activated" spectrum (Gordon 2003; Wynn et al. 2013)). Such abnormalities can contribute to explain the triggering of cancer, and they could be targeted by a number of therapies, which precisely aim at correcting these immune defects.

In the second situation, however, the immune system acts normally and immune-mediated decohesion is due to an abnormal context. Pathogens,

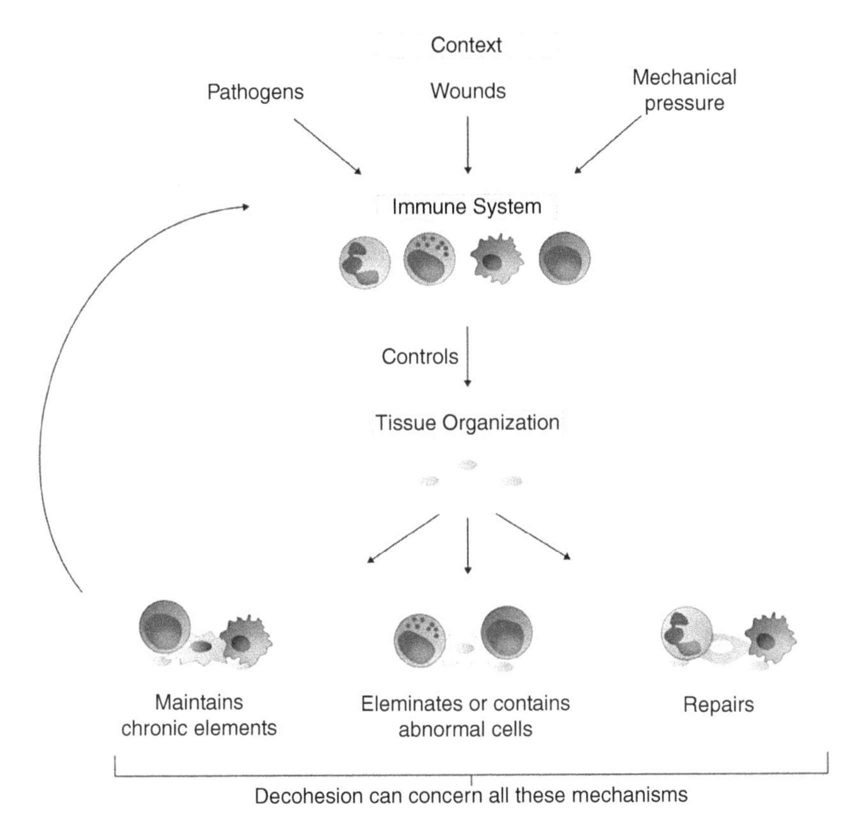

Figure 4.4 Decohesion in cancer induced by an abnormal context. Immune-mediated decohesion may be due to an abnormal context rather than an abnormal immune system. This abnormal context can be due to the presence of pathogens, wounds, mechanical pressures, and carcinogens of environmental origins, among many other resources; it can also be triggered by the tumor itself. In many situations, the decohesion mediated by the immune system results from abnormal realization of normal processes (such as maintenance and repair). (Figure drawn by Wiebke Bretting).

wounds, mechanical pressure, and local modifications due to carcinogenic environmental factors, among several other causes, can create an abnormal local context (characterized by inflammation, perturbation of the extracellular matrix, and so on) (Dolberg et al. 1985; Bissell and Radisky 2001; Mantovani et al. 2008; Fernández-Sánchez et al. 2015). This local context influences the immune system, which in turn responds as it usually does, that is, by maintaining or repairing the tissue – even if the final, pathological, outcome is cancer promotion (Figure 4.4). It has long been known, for instance, that tumors resemble "wounds that do not heal" (Dvorak 1986, 2015; Schäfer and Werner 2008), which means that, in cancer, normal repair mechanisms are triggered but generally without reaching the "resolution phase" (which, in the physiological context, is indispensable to terminate the reparative process (Gurtner et al. 2008)). Furthermore, the tumor itself can be a major source of perturbation of the local context: it can influence the immune system through a variety of cytokines and can also increase inflammation and wounding, modify blood vessels, reshape the extracellular matrix, or exert a mechanical pressure, among many other possibilities. This is often described as the "hijacking" or "co-option" by the tumor of physiological pathways and of the tissue microenvironment (Kitano 2004a; Greaves 2007; Lean and Plutynski 2016). Despite its importance, one must keep in mind that such co-option is only one of the many ways in which the local context can become abnormal and favor the contribution of immune processes to cancer progression.

In all these contexts, immune-mediated decohesion results from the abnormal realization of normal processes, and this can help explain why tumors largely resemble organs (Egeblad et al. 2010) and are the products of classic developmental (da Costa 2001; Radisky et al. 2001; Huang et al. 2009) and reparative pathways (Bissell and Radisky 2001) realized in an abnormal context. There is no doubt that the outcome is pathological (the promotion of a cancerous tumor, including in some cases metastatic spread), but the immune system, in many of these circumstances, does not strictly speaking "dysfunction"; it just does what it always does (maintaining the local environment, repairing in case of wound, etc.).

If the view presented here is correct, then a much richer picture emerges about how the immune system influences cancer and, ultimately, of potential therapeutic opportunities as well. Indeed, the immune system influences cancer through different processes (elimination, containment, maintenance, repair, and so on), via many actors (not only lymphocytes, but also macrophages, neutrophils, and various cytokines), at several different levels (within the tumor, but also around the tumor, in the whole tissue, and at a systemic level in the organism), and at all temporal stages of cancer progression (initiation, neoplastic progression, and metastasis). Additionally, the influence of the immune system on the cancerous tumor can be negative (the immune system prevents cancer progression) or positive (the immune system promotes cancer progression). All this suggests a whole series of new opportunities for investigating immunotherapies, which could, at least in principle, target these different processes, actors, levels, and temporal stages. Current immunotherapies (particularly immune checkpoint inhibitors and CAR-T cells (Joyce and Fearon 2015; June and Sadelain 2018)) focus on lymphocytes in terms of actors and on elimination and maintenance and rupture of chronicity in terms of processes, but many other possibilities exist. Depending on where we are in the cycle of cancer–immune system interactions and on the actors of the tumor microenvironment involved, some therapeutic strategies will aim at normalization while others will aim at denormalization. Examples of normalization include the reduction of the level of inflammation in the tissue, the elimination of pathogens and/or chronic wounds, the restoration of immune accessibility to the tumor, and the facilitation of the resolving phase of tissue repair. In contrast, denormalization would be a major aim when the immune system interacts with tumor components as if they were normal constituents of the body, as, for example, when the immune system is tolerogenic in the context of chronically present tumor antigens or when the immune system continuously triggers repair pathways to respond to a local cancerous context that displays many features usually associated with a wound.

In summary, this section has shown that focusing on the immune system is essential for anyone studying cancer. Cancer is a disease of multicellularity and, more specifically, of the *cohesion* of the multicellular organism. Immunological surveillance constitutes one of the main and best described mechanisms by which the multicellular organism exerts control over lower-level entities. A major result of recent research is that the immune system can both restrain and promote cancerous tumors, which may seem, at first sight, paradoxical. Yet the situation becomes less paradoxical when one realizes that immune-mediated decohesion is often due to an abnormal context rather than a dysfunctional immune system. We have suggested here an extended view of cancer–immune interactions, which opens up many opportunities for investigating new mechanisms of tumor control and tumor promotion and, ultimately, for developing novel therapeutic opportunities based on the action of the immune system.

5 Neuroimmunology: The Intimate Dialogue between the Nervous System and the Immune System

> Whatever forces were operating to set immunology apart, recent data suggest that much could be learned by studying immunoregulation as part of an integrated network of adaptive processes including behaviour. If not now, when? (Ader and Cohen 1985)
>
> These exciting revelations place neuroimmunology at the forefront of biomedical research priorities. With the potential to affect such a diverse array of neurological ailments, many of which have no known therapy, the hope is that an improved understanding of immune-CNS interactions will bring to light new paradigms for preventing and treating neurological disease. (Mueller et al. 2016)
>
> These findings suggest that the brain–cytokine system, which is in essence a diffuse system, is the unsuspected conductor of the ensemble of neuronal circuits

> and neurotransmitters that organize physiological and pathological behavior. (Dantzer et al. 2008)

If you fall asleep reading this Element, this is probably simply because you are tired or because its content is not entertaining enough. But if you are literally *falling asleep*, then perhaps you have narcolepsy, a disorder estimated to affect around one in 2,000 people. The symptoms of narcolepsy, which usually begin in adolescence or early adulthood, include daytime sleepiness and, in some cases, cataplexy – sudden muscle weakness during wakefulness that causes falls. Severe forms of narcolepsy are associated with abnormally low numbers of neurons that produce hypocretin, a protein that controls sleep–wake cycles. Recently, it was suggested that narcolepsy might be the consequence of an autoimmune response (following older work on association with some HLA alleles). Narcoleptic patients have immune $CD4^+$ memory T cells that target peptide fragments of hypocretin, suggesting that autoimmunity could play a role in narcolepsy, although the exact causal relationships remain to be determined (Liblau 2018).

A problem that affects many more people than narcolepsy is depression. All readers of this Element certainly know somebody who has depression and/or have experienced depression themselves. Depressive disorder, a multiform and multifactorial condition, was estimated to affect 8.5 percent of people in Europe (Ayuso-Mateos et al. 2001). For a long time, depression has been considered to be a psychiatric disease, usually treated with serotonin-tweaking drugs like Prozac. Yet a growing number of researchers look at depression from an additional perspective based on immunology (Bullmore 2018; Dantzer 2018) (more about this example below).

Such examples and many others suggest that the nervous system and the immune system, far from being separated, can interact intimately in almost all metazoans, including humans. The study of these interactions has given rise to an interdisciplinary domain, neuroimmunology. My main objective in this section is very modest: it is to offer a conceptual clarification of the different issues raised by neuroimmunology, which often remain intertwined and insufficiently distinguished. To do so, after a short history of neuroimmunology, I present important results concerning the interactions between the nervous and the immune systems in health and disease. I then propose to distinguish five different conceptual questions when dealing with neuroimmune interactions, and finally I mention some important philosophical consequences neuroimmunology can have, particularly about cognition. So, what will be said in this section is no more than a preliminary conceptual and philosophical exploration of the field of neuroimmunology. My hope is simply to convince some readers to take up the challenge of immersing themselves in this fascinating domain.

5.1 From Psychoneuroimmunology and Neuroimmunology to Present-Day Characterizations of the Dialogue between the Nervous and the Immune Systems

Despite important preexisting research on interactions between the nervous and the immune systems (relative to the blood–brain barrier, the idea of the brain as an immune-privileged organ, and the role of microglia cells, in particular; more will be said below about all these aspects), the fields of neuroimmunology and psychoneuroimmunology emerged in the 1970s and 1980s. The Neuroimmunology Branch of the National Institute of Neurological Disorders and Stroke was established in 1975 (McFarland et al. 2017). The journal *Neuroimmunology* was founded in 1981. Terms such as neuroinflammatory were already in use in the early 1980s (e.g., (Hartung and Toyka 1983)). Psychoneuroimmunology was also born in the second half of the 1970s, following the work of Robert Ader (1932–2011) and a few others. The first edition of the volume *Psychoneuroimmunology*, edited by Ader, was published in 1981 (Ader 1981) and several updated editions followed. In 1987 Ader founded the journal *Brain, Behavior, and Immunity*, the official journal of the Psychoneuroimmunology Research Society. Importantly, the domain of psychoneuroimmunology stemmed in part from psychosomatic medicine (Kiecolt-Glaser et al. 2002).[1]

Despite considerable overlap between neuroimmunology and psychoneuroimmunology, neuroimmunology tends to investigate the interactions between the nervous and the immune systems, particularly at the cellular and molecular levels, while psychoneuroimmunology explores how behavior influences and is influenced by the immune system. Historically, psychoneuroimmunology has focused on the conditioning of immune responses, the role of immune factors in mental disorders such as schizophrenia, and the effect of psychological factors such as stress on immune responses (Ader 2000). Later, the role of neuroimmune interactions in fatigue and sickness behavior (the feeling of fever and nausea, for example, when we are sick, and which is mediated by pro-inflammatory cytokines such as IL-1β) became important topics in the field (Dantzer et al. 2008). As early as the 1970s, interactions under investigation included not just two but three actors: the nervous and the immune system, but also the endocrine system (Besedovsky and Sorkin 1977) (for a retrospective, see (Besedovsky and Rey 2007)). This explains why some researchers talk about "psycho-neuro-endocrino-immunology" (Sivik et al. 2002). Neuroimmunology, psychoneuroimmunology, and psycho-neuro-endocrino-immunology all pay attention to both the central nervous system (CNS) and the

[1] I thank Jan Pieter Konsman for many discussions about the history of neuroimmunology and psychoneuroimmunology.

peripheral nervous system (PNS, the part of the nervous system that is outside the brain and spinal cord).

Strikingly, the fields of neuroimmunology and psychoneuroimmunology, since their inception, have given rise to controversies and have often been accused of not following the most rigorous scientific standards. This accusation has particularly targeted psychoneuroimmunology, as illustrated, for instance, by the disagreement between (Maddox 1984) and (Ader and Cohen 1985), and as discussed in detail by (Cohen 2006).

Here we will use neuroimmunology in an inclusive sense to refer to all the approaches nowadays that study interactions between the nervous and the immune system, at all levels, in health and disease, and with or without a focus on behavior.

5.2 Interactions between the Nervous and the Immune System in Health

Until the 1990s, a widespread conviction was that the brain was an "immuno-privileged" organ (Carson et al. 2006), understood as the idea that the brain is devoid of immune cells. The brain and the immune system were thought to be separated by a strict barrier, the *blood–brain barrier* (made of tightly packed endothelial cells, restricting the passage of many substances into the parenchyma), and the crossing of this barrier by immune cells was thought to be pathological and dangerous. Yet as with all other organs thought to be immuno-privileged (see Section 3), recent research has emphasized first that the brain has its own immune system, and second that the blood–brain barrier is partially permeable.

The main cellular actors of the brain immune system are microglia, the resident immune phagocytes of the CNS (see Figure 5.1). They constitute about 10 percent of the total cells in the adult CNS. Although microglia were documented as early as the 1920s, the conception of their roles and functioning has been considerably extended during the last decade (Salter and Beggs 2014). In the healthy CNS, microglia, long conceived as dormant, are in fact highly active, continuously monitoring their environment with extremely motile processes and protrusions (Nimmerjahn et al. 2005) and interacting with neurons and other brain cells. Microglia initially were thought to simply react to CNS injury, infection, or pathology, but recent work suggests that they play a key role in synaptic remodeling both in development and in adult life, refining neuronal circuitry and network connectivity, and contributing to neuronal plasticity (Wu et al. 2015). Importantly, the immune complement (Stephan et al. 2012) and the major histocompatibility complex class I molecule H2-D^b (Lee et al. 2014) are also instrumental in CNS synapse pruning. Overall, microglia constitute a perfect illustration of what we described in Section 2, namely the fact that

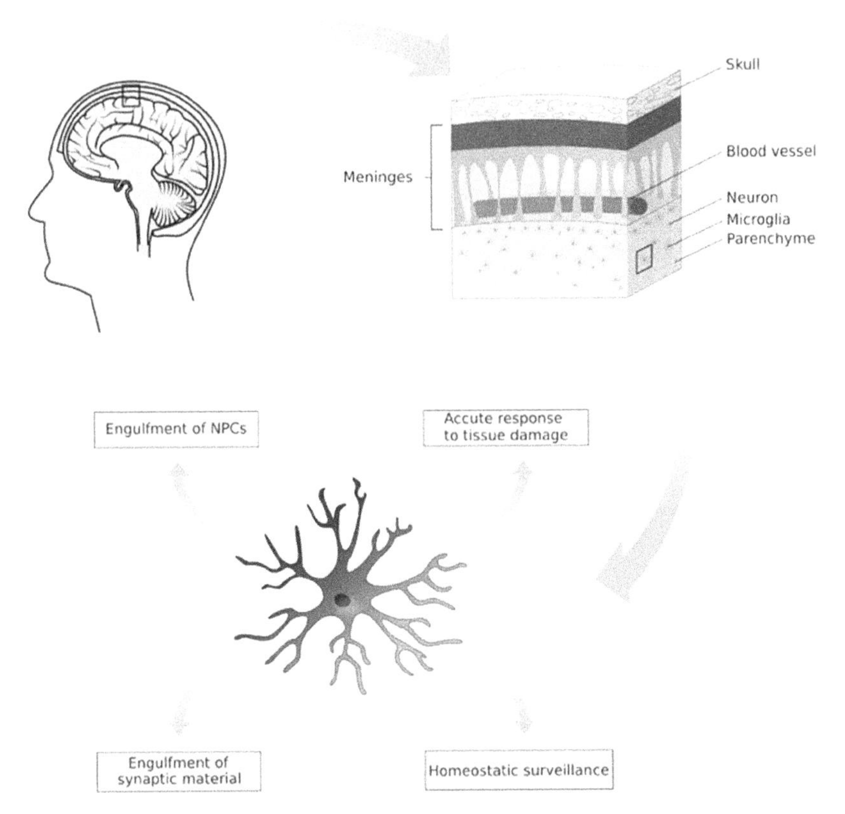

Figure 5.1 Microglia and their many activities. Microglia are a major element of the brain's immune system. These resident immune phagocytes constantly monitor their microenvironment and participate in many processes in health and disease, including engulfment of neural progenitor cells (NPCs), acute response to CNS damage, engulfment of synaptic material, and homeostatic surveillance. (Figure drawn by Wiebke Bretting).

the activities of immune cells include but are not limited to defense, comprising also development, repair, clearance of debris, and so on (Michell-Robinson et al. 2015).

The long-dominant view has been that, in healthy conditions, the brain is devoid of lymphocytes, as these cells could cause major damage. Recently, however, a "peri-cerebral" adaptive immune system has been described (Figure 5.2): the meninges contain lymphatic vessels that remove waste from the parenchyma, can relay information about possible infections in the brain, but also harbor peripheral immune cells that communicate with the brain via cytokines (Louveau et al. 2015b; Kipnis 2016). This discovery confirms that

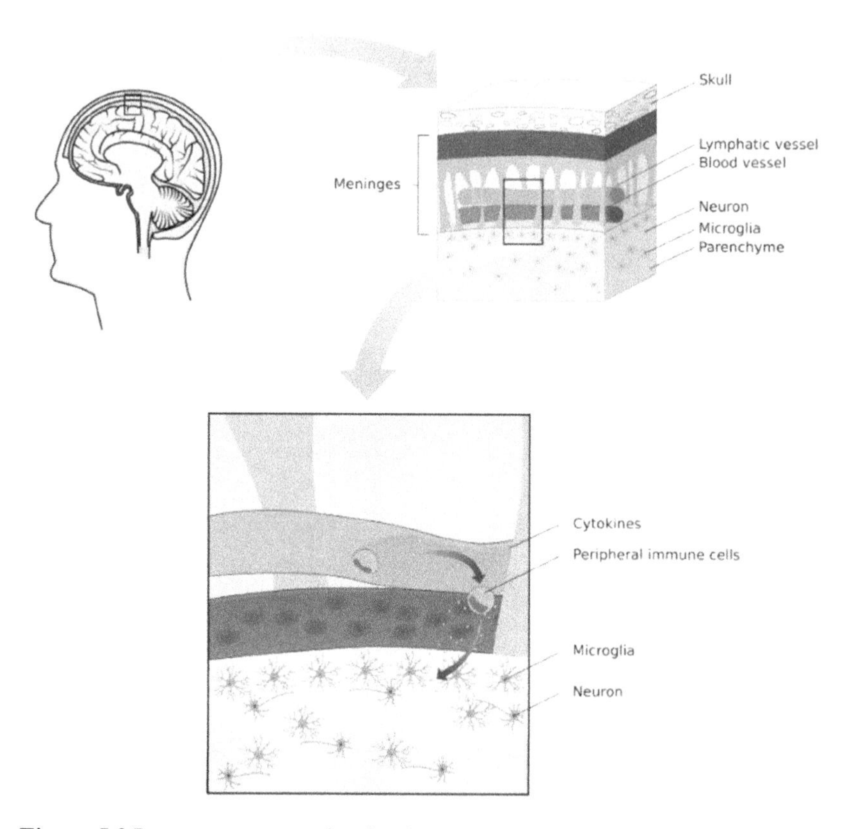

Figure 5.2 Immune communication between meninges and the brain. Apart from microglia, immune cells are generally not present in the brain, as, in nonpathological situations, they are thought to not cross the blood–brain barrier. However, meninges contain lymphatic vessels and peripheral immune cells, which communicate with the brain via cytokines. (Figure drawn by Wiebke Bretting).

there is a rich immunological crosstalk between the brain and the rest of the body, particularly via the meninges.

Collectively, these data confirm that there is a cellular and molecular immune system of the brain. The actors and mechanisms just described come in addition to various cell-intrinsic innate defense mechanisms used by neurons in case of viral infections (Ordovas-Montanes et al. 2015).

Not only is there an immune system of the brain, but recent research confirms that there are many interactions between the nervous and the immune system in a healthy organism, with much richer communication pathways than initially suspected (Figure 5.3). These interactions occur at various levels: molecules, cells, organs, and at the systemic level. There is a dense sympathetic innervation of all lymphoid organs (Dantzer 2018). The PNS regulates immunological development (the sympathetic nervous system regulates haematopoiesis), priming (neurons influence the triggering of an immune response in lymph nodes), and deployment (peripheral neurons associated with vessels can impact on leukocyte recruitment into peripheral tissues) (Ordovas-Montanes et al. 2015). In response to pathogens or tissue perturbation, immune cells are activated at the periphery and release cytokines and other inflammatory molecules; these molecules have an impact on local sensory neurons and influence signaling to the CNS (Chavan et al. 2017). Furthermore, pro-inflammatory cytokines produced by immune cells at the periphery communicate with the brain through afferent nerves, a process that leads to the production (by activated microglia) of other pro-inflammatory mediators in the brain itself. (Using a perhaps slippery vocabulary, some authors say that the brain forms, via neuromediators and immune mediators, an "image" of immune responses occurring in peripheral tissues (Dantzer et al. 2008)).

Interactions between the nervous and the immune system can have important functional consequences. For example, homeostatic circuits regulating temperature maintenance, blood pressure, and intestinal mobility involve immune cells. The vagus nerve is important for detecting and reporting on peripheral immune responses and, in turn, efferent signals from the CNS are indispensable for the regulation of inflammatory responses (Ordovas-Montanes et al. 2015). Kevin Tracey in the 2000s proposed calling this neuroimmune network the "inflammatory reflex" (reviewed in (Chavan et al. 2017)), a concept enriched and discussed in subsequent research (Dantzer 2018).

Neuroimmune interactions can also involve additional actors. A major recent example is research on the microbiome–gut–brain axis. Mouse and insect models suggest that the microbiome influences brain development and behavior (Sharon et al. 2016; Vuong et al. 2017; Schretter et al. 2018b), in part through the mediation of the immune system (Fung et al. 2017). Whether this conclusion may apply to humans remains an open question.

Crucially, neuroimmune interactions have recently been said to have an impact on cognition. It has been proposed that cytokines play a critical role in spatial memory (Sparkman et al. 2006) and that microglia are important for learning and memory by promoting learning-related synapse formation through brain-derived neurotrophic factor signaling (Parkhurst et al. 2013). Furthermore, according to some authors, adaptive immunity influences cognition (Kipnis 2016). Mice deficient in T lymphocytes were found to exhibit cognitive impairment in spatial learning/memory tasks and passive transfer of mature T cells improves their cognitive function (Kipnis et al. 2004). A likely

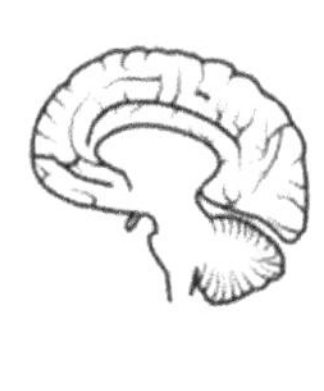

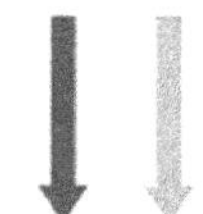

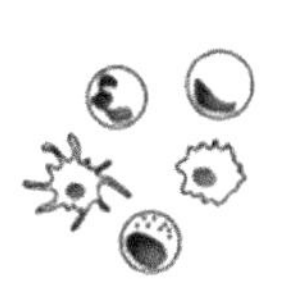

Figure 5.3 Communication between the nervous and the immune system. According to the initial view (top) neuroimmune interactions were thought to occur via neuroendocrine mediators released in the general circulation. According to the current view, however (down), long-distance interactions between the nervous and the immune system are mediated by neural pathways much more than by circulating neuroendocrine mediators, and the communication is bidirectional. (Figure drawn by Wiebke Bretting, after Dantzer 2018).

hypothesis is that pro-cognitive T cells are autoimmune cells stimulated by brain-derived molecular cues (Kipnis et al. 2012). The influence of T cells on cognition could be mediated by interleukin-4 (IL-4), as the learning behavior of mice that do not express IL-4 is substantially lower than that of wild mice, and

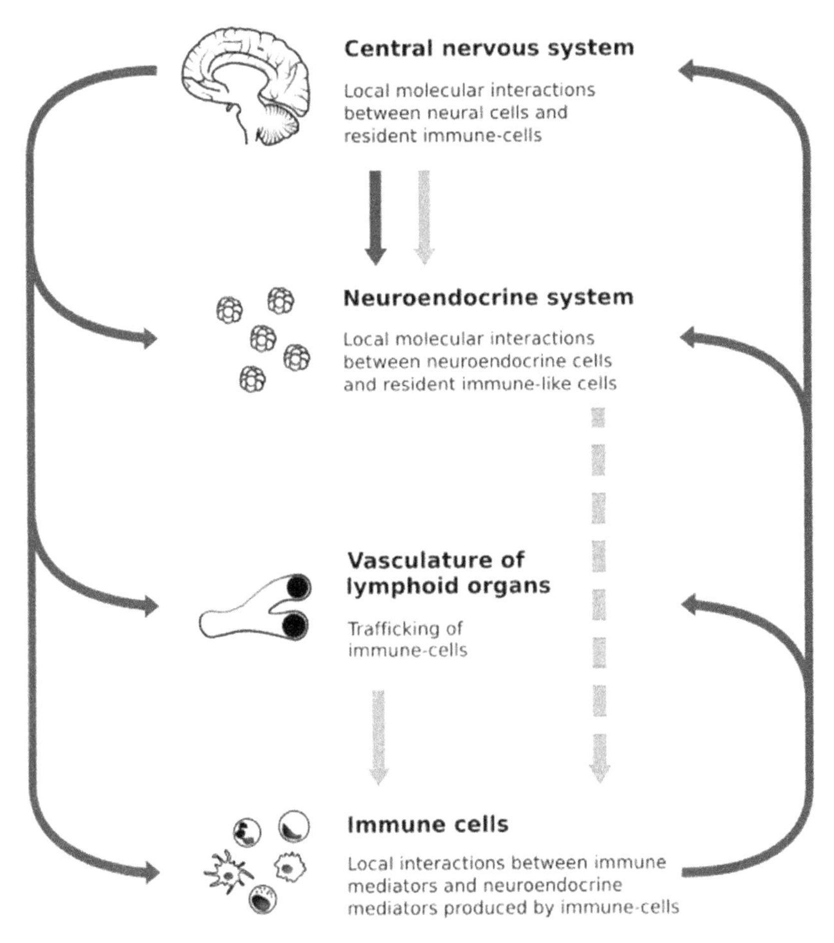

Figure 5.3 (Cont.)

this effect can be reversed by injecting wild-type T cells but not T cells that do not express IL-4 (Derecki et al. 2010). Another, nonexclusive, possibility is that this effect is mediated by myeloid cells. More recently, it has been claimed that mice deficient in adaptive immunity exhibit social deficits and hyperconnectivity of fronto-cortical brain regions, that social deficits are reversible via repopulation with lymphocytes, and that interferon-γ is a probable molecular link between meningeal immunity and neural circuits involved in social behavior (Filiano et al. 2016). Further work is needed to unravel the exact molecular and cellular mechanisms by which neuroimmune interactions affect cognition, but data accumulated in the last decade is highly promising and may open up new important avenues for research.

In summary, research in neuroimmunology over the last four decades has shown that there are many interactions between the nervous and the immune system. Neuroimmune interactions can be studied at different levels in the organism (from molecules to systems and even with an attention to environmental influences), at different scales (some neuroimmune communication pathways are short-distance, others are long-distance), and through an attention to different routes (direct interactions, but also indirect interactions, such as those involving endocrine elements or the microbiota).

5.3 Interactions between the Nervous and the Immune System in Pathological Contexts

According to some authors, neuroimmune interactions play a role not only in physiological but also in pathological contexts. Many disorders classically defined as neuronal and/or psychiatric are now said to possess a significant immunological component (Kipnis 2016). This is the case, for example, for some neurodegenerative disorders (Heneka et al. 2014) such as multiple sclerosis, which is an autoimmune disease (Dendrou et al. 2015), but also in Alzheimer's disease (Heppner et al. 2015). In addition, several lines of evidence suggest that various psychiatric pathologies could be caused, at least in part, by components of the immune system. Some authors emphasize that antineuronal autoantibodies are found in the serum of 11.6 percent of 925 patients admitted to acute psychiatric inpatient care (Schou et al. 2016), others that 20 to 40 percent of acute psychiatric inpatients exhibit low-grade inflammation, suggesting the possibility that inflammation can be relevant for many psychiatric disorders (Osimo et al. 2018) (as typically recognized for schizophrenia (Khandaker et al. 2015)). Many of these observations are limited and only correlative, so more work is needed to draw any conclusion here. An additional layer of complexity is that some autoimmune diseases such as lupus are accompanied with neuropsychiatric manifestations, which may be due to the existence of brain-reactive autoantibodies (Williams et al. 2010).

Several researchers propose that the immune system plays an important role in major depressive disorder. In humans, one-third of patients treated with cytokine therapies (IL-2 and IFN-α) for cancer or chronic viral hepatitis develop major depressive disorders (Raison et al. 2006). Clinical observations, epidemiological studies, and investigations in animal models have tended to confirm that pro-inflammatory cytokines can induce depressive-like behaviors (Dantzer et al. 2008). All this has contributed to the development of a non–brain-centric view of depression. The causal involvement of indoleamine 2,3-dioxygenase (IDO), an immune-inducible enzyme that metabolizes tryptophan along the kynurenine pathway and plays a major role in immunoregulation, was shown in animal models. Furthermore, pharmacological or genetic blockade of IDO activation abrogated depression-like behavior in mice (Dantzer 2018).

Collectively, these investigations on the role of neuroimmune interactions in psychiatric disorders have led to the emerging field of immunopsychiatry (Pariante 2015). This domain has its origins in neuroimmunology and psychoneuroimmunology but focuses on the influence of immune mechanisms on behavior rather than the other way around.

If confirmed by future results, research on the role of the immune system in neurological and psychiatric diseases could have interesting and innovative therapeutic consequences. It has the potential to expand traditional approaches that see such diseases as purely mental and/or neurological, especially if future investigations confirm that some of these diseases can be managed via targeting of the immune system (among other components). This includes major depressive disorder (Miller and Raison 2016; Bullmore 2018), bipolar disorder, schizophrenia, and several others (Miller and Buckley 2017). The possibility to treat neurodegenerative disorders via immunotherapies also is under intense investigation (Weiner and Frenkel 2006).

Interestingly, an immunological approach could lead to redefined nomenclatures of psychiatric disorders and treatments. Diagnosis for psychiatric disorders is currently based on descriptive nomenclature because of a lack of clearly defined causal mechanisms. Yet some immunological mechanisms are common to psychiatric disorders belonging to different categories, hence the suggestion to develop alternative nomenclatures based on immunological characteristics as well as associated environmental factors (Leboyer et al. 2016).

Reciprocally, some diseases traditionally classified as immune may be targeted by acting on the nervous system. For example, in a recent clinical study, electrical vagus nerve stimulation was successfully used to improve symptoms in rheumatoid arthritis, a chronic inflammatory and autoimmune disease (Koopman et al. 2016). Further investigations are needed, but some specialists consider that bioelectronic devices can be used to modulate neural circuitries, constituting a complement to drug treatment, especially in immune diseases (Chavan et al. 2017)

The field of immunotherapy-based approaches to neurological and psychiatric diseases is still in its infancy. Yet if successful, it could contribute to closing the gap between psychiatry and the rest of medicine, and it could lead to a radical change in the way we traditionally conceive mental health problems.

5.4 Mapping the Different Conceptual Questions Raised by Neuroimmunology

Recent scientific literature abounds in bold claims about the relation between the nervous and the immune system. These two systems are often said to have strong interactions and many similarities, perhaps a common evolutionary origin, while some suggest that they overlap to a great extent and even that they could constitute a single system (see details below). I suggest that, even if several of such claims are often formulated as if they were intimately connected or even equivalent, it is crucial, for the sake of clarity, to disentangle these claims. My analysis of this literature has led me to single out five dimensions of neuroimmunology, each corresponding to a different question (Figure 5.4): interaction, similarity, overlap, origins, and control. Importantly, distinguishing five questions in neuroimmunology can serve not only as a mapping of the various goals currently pursued by researchers in this domain but also as an invitation to do further investigations to better address some questions that have tended to remain in the background in the last two decades.

5.4.1 Interaction: How Do the Nervous and the Immune System Interact?

The question of how exactly the nervous and the immune system interact is the most basic and the most extensively discussed in neuroimmunology. Some of the main results of this research have been presented in the previous two sections. My aim here is not to give more details about this question but rather to explain why it should not be confused with the four other questions that are often raised in the scientific literature.

5.4.2 Similarity: Are the Nervous and the Immune System Structurally and/or Functionally Similar?

The nervous and the immune systems are often said to be similar, that is, to share several important features. Similarity can be at a structural level. Both systems communicate via soluble ligands and receptors. Molecules that mediate such communication include cytokines, chemokines, neuropeptides, neurotransmitters, neurotrophins (Camacho-Arroyo et al. 2009), and their receptors. Perhaps more distinctively, both systems make use of specific structures called synapses. The term appeared in neurobiology in the late nineteenth century and was subsequently adopted by immunologists in the 1980s to describe the extended communication surface platforms established between two immune cells, particularly antigen-presenting cells and lymphocytes (Steinman 2004). Both types of synapses are stable adhesive junctions

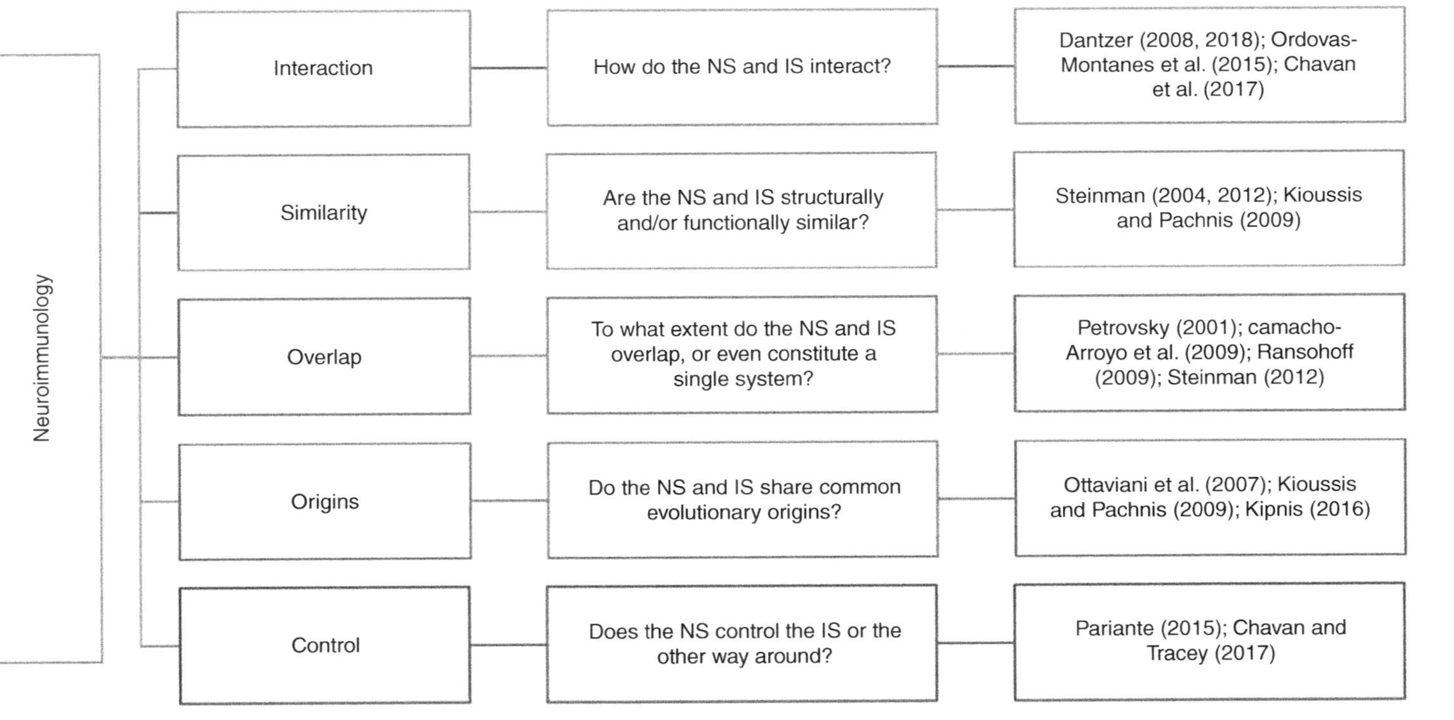

Figure 5.4 Neuroimmunology: A conceptual tree. This figure distinguishes five different questions raised by current scientific literature in neuroimmunology, and which in general are insufficiently separated.

between two cells across which information is transmitted via secretory molecules. The detailed resemblances and differences between the neural and immunological synapses are an important object of study (Dustin and Colman 2002).

The similarity can also be at the functional level. It is often said, for example, that both systems recognize their target and display a form of memory insofar as they respond differently to a second stimulation by a given stimulus, as if they could remember a past experience (Kioussis and Pachnis 2009). Sometimes, this functional comparison goes very far, as when it is suggested that the immune system is, in itself, "cognitive" (Tauber 1997). The exact meaning of this claim is unclear. It dates back to at least Niels Jerne (Jerne 1985). More recently Irun Cohen (Cohen 2000, pp. 181–189) defended the view that the immune system perceives signals and responds to them by a decision-making process, itself based on a complex language and a capacity to remember (immunological memory). It is uncertain how comparisons made at such a level of generality may prove scientifically useful because they are difficult to operationalize and test. They could perhaps generate novel avenues for research, but they may also be experimentally sterile, as happened to a large extent with Jerne's views.

Other functional similarities often mentioned include plasticity and motility (Kioussis and Pachnis 2009). Both neural and immune cells are plastic to a high degree: depending on microenvironmental cues, they can differentiate into phenotypically distinct subtypes. Moreover, cells of both systems are able to carry information from and to distant parts of the body. Yet as we will see, they use different means to do so.

Overall, similarities between the nervous and immune systems are numerous and often informative. Moreover, there are many promising avenues for future research in this area, including, for instance, a comparison between the processes of epigenetic regulation in neural and immune cells.

Neuroimmunologists often switch from the idea that the nervous and the immune system interact to the idea that the two systems are similar, or the other way around (e.g., Steinman 2012; Ordovas-Montanes et al. 2015). It should be clear, though, that interaction and similarity are two different things. Two entities can have strong interactions without being similar, and it is perfectly possible that two similar things have no interaction at all. As self-evident as such distinctions might seem, they do point to some confusions that exist in the scientific literature. For example, even if the immunological and the neural synapses do not interact, it remains extremely interesting to study their similarities. And even if an immune cell at the periphery and a neuron in the brain are structurally and functionally dissimilar, understanding how they interact during an infection or an autoimmune disease remains crucial. It is reasonable for future neuroimmunological research, therefore, to study both similarities and interactions even when they do not go hand-in-hand.

5.4.3 Overlap: To What Extent Do the Nervous and the Immune System Overlap or Even Constitute a Single System?

Sometimes, comparisons between the nervous and the immune system go beyond mere similarity, especially when functional features considered as typical of one system are exhibited by the other system. A key discovery was that immune cells express receptors for neuromodulators and neurons express immune receptors, which is crucial for their capacity to influence each other. Cytokines, classically attributed to the immune system, are also produced by cells of the CNS (microglia, astrocytes, and neurons), where they regulate the development of the nervous system as well as some of its most crucial physiological processes, including neurotransmission (Camacho-Arroyo et al. 2009; Ransohoff 2009). Conversely, neuropeptides, long thought to be specific to the nervous system, are now known to be produced by immune cells as well (Steinman 2004). Because of this overlap, traditionally conceived functions are often blurred: elements classically defined as pertaining to the nervous system play important immunological roles and vice versa.

Neuroimmunologists frequently put together the "interaction" question and the "overlap" question (Ordovas-Montanes et al. 2015; Chavan et al. 2017). However, overlap does not always come with interaction, as illustrated by many cases of functional redundancy (where identical or nearly identical components can realize a given function) and functional diversity (where heterogeneous components can realize a given function) in studies on robustness (Kitano 2004b). Two engines in an airplane overlap functionally, but it is preferable that they don't have too strong an interaction, otherwise damage to one motor might disrupt the other. Such fail-safe mechanisms abound in biology. Reciprocally, interaction does not necessarily imply an overlap. If a cytokine produced by the immune system interacts with the nervous system, it is important to study it, even if it is not produced by the nervous system as well.

Furthermore, scientists enthusiastic about functional convergences between the nervous and the immune system tend to switch from the idea of overlap to the idea of unification, that is, the idea that the nervous and the immune system (together with the endocrine system, in general) constitute one single system (Petrovsky 2001; Steinman 2012). But here too, caution is in order: clearly two systems can overlap without constituting a single system. This is illustrated, again, by studies on biological robustness. Many cases of functional redundancy and functional diversity rest on the fact that two systems realize entirely or partially similar functions without being one and the same system because in those cases distributivity is essential for maintaining robustness. For example, the two systems will not be activated in the same circumstances or at the same time, and/or the failure or insufficiencies of one system can be compensated by the other. One promising research program for future neuroimmunology is to outline and explain the circumstances in which the nervous and the immune system really overlap in space and time, those in which they act at different moments or different places, as well as those in which one system compensates for the other or takes over from the other.

5.4.4 Origins: Do the Nervous and the Immune System Share Evolutionary Origins?

When discussing interactions, similarities, and/or overlap between the nervous and the immune system, many neuroimmunologists make claims about their supposed common origins in evolution. Such claims are often made in general terms (Kipnis 2016; Veiga-Fernandes and Pachnis 2017), but sometimes they are more specific, as when it is suggested that a common origin is a likely explanation for the fact that the two systems display a common capacity for monitoring and responding to changes in the external and internal environments (Kioussis and Pachnis 2009). Based on various techniques, including immunochemistry and sequence analysis, Ottaviani and colleagues (Ottaviani et al. 2007) defended the view that there was a common evolutionary origin for the immune and neuroendocrine systems.

As stimulating as these claims may be, it is important to remain careful about them. First, the comparative study of the evolutionary history of the nervous and immune systems is still in its infancy. Biologists often act as if both systems appeared approximately at the same time, but this is clearly not the case. Immune systems are much more widespread than nervous systems in the living world and much older in life's history. Plants and prokaryotes have an immune system but they don't have a nervous system. Metazoans all have an immune system but not all have a nervous system. Sponges, which have an immune system (Müller 2003), are generally considered to not have a nervous system, even though they may possess neuro-sensory-like cells (Miller 2009). And if one decides to focus on adaptive immune systems (as done by some neuroimmunologists), then the reverse is true: countless animals, such as arthropods, for example, possess a nervous system without having an adaptive immune system. This is all the more important as, although neuroimmunologists have worked mainly on mammals, intimate neuroimmune interactions have been found across the animal kingdom. For example, recent work has shown that IL-17, a pro-inflammatory cytokine, is a neuromodulator and contributes to behavior in *Caenorhabditis elegans* (Chen et al. 2017). Thus, much more research is needed about the history of the nervous and immune systems in all their diversity before making a claim about their possible common origins.

Second, one cannot make direct inferences from similarity to common evolutionary origins. There are obviously many different possible evolutionary explanations for similarities, including homology (shared ancestry) and analogy (convergent evolution). In future research deciding between these two options at the system level will be important but it will also be difficult. The question has also been raised at the cell level: did neural and immune cells evolve independently but later co-opt functions from each other, or did they evolve from a common ancestral cell able to recognize and interpret the environment, communicate with other cells, and exhibit plasticity (Kioussis and Pachnis 2009)?

Additionally, evolutionary explanations will differ depending on whether we want to explain interactions, similarities, or overlap between the nervous and the immune system. This is more confirmation that one should be careful when switching from one question to the other.

So, exciting research remains to be done for scientists and philosophers interested in when and why nervous and immune (sub)systems emerged in evolution and how they fit together. An interesting proposal has been made on this topic by my colleague Jean-François Moreau (personal discussions). In his view, the functioning and origins of the nervous and the immune system must be put into the context of the emergence of multicellularity. Multicellularity presupposes internal communication and in metazoans three types of long-distance communication channels can be distinguished. The first resembles our electric networks; it corresponds to the nervous system, and more specifically to neurons, which can send information at a very high speed with a relatively diverse content. The second resembles our water and/or sewage networks; it corresponds to the blood and lymphatic vessels, where endocrine signals, in particular, circulate. It delivers information at a relatively high speed with a relatively diverse content. These first two systems (nervous and vascular) are rigid: they can be modified (via neurogenesis and angiogenesis, for example) but only at an extremely slow rate. The third system resembles our mail carriers; it corresponds to immune cells, which are the uniquely mobile cells of the organism and which can deliver information with extremely diverse content everywhere in the organism, often over long distances though at a limited speed. In addition to carrying information, immune cells can perform all sorts of activities, including pathogen clearance, tissue remodeling, and tissue repair, among many others (Eom and Parichy 2017). Overall, this view suggests interesting and important distinctions about the different selective pressures that might have existed at the origins of nervous, endocrine, and immune systems in metazoans. It is likely that some types of messages can be delivered by one system only: for example, high-speed communication is best realized by the nervous network, while other processes such as pathogen clearance and tissue repair require the unique mobility of immune cells. All this constitutes an invitation to explore not only the similarities between the nervous, immune, and endocrine systems but also their complementarities, as well as the means that evolved to coordinate these systems characterized by their different communication channels. Among other benefits, such research could constitute an important contribution to current scientific and philosophical discussions about how and why nervous systems originated in the animal world (Miller 2009; Keijzer et al. 2013; Godfrey-Smith 2016) by enriching the context of this question with considerations about possible complementarities (and possible trade-offs as well) between the nervous system and other bodily systems.

5.4.5 Control: Does the Nervous System Control the Immune System or the Other Way Around?

A question slightly different from the four others yet important in this discussion is control. Many neuroimmunologists switch from the description of intimate interactions between the nervous and the immune system to the idea that one system controls the other (Chavan and Tracey 2017). When describing what he sees as the emerging field of immunopsychiatry, Pariante (2015) suggests that this domain, by giving prominence to the immune system, reverses what "governs" and "is governed."

It should be clear, however, that interactions between two systems and control by one system over another are two entirely different things. It is not because recent research has shown that the immune system could influence the nervous system and its behavior that we should conclude that the immune system controls the nervous system let alone behavior. More data would be needed to demonstrate control though it is not entirely clear which data could be considered conclusive in that case. Moreover, the choice between the "controller" and the "controlled" seems to be more dependent on the disciplinary background of the person making the claim than on anything else. Personally, I don't see the need to attribute control to one system or another.

In summary, we have singled out five questions in the neuroimmunological literature and suggested that distinguishing them was useful not only as a conceptual clarification of current research in that domain but also as an invitation to further explore the questions that, as is typically the case with that of evolutionary origins, have tended to remain in the background because their specificities have not been sufficiently recognized.

5.5 Conclusion: Some Philosophical Consequences

Let's end this section by drawing three philosophical consequences from this exploration of neuroimmunology. First, neuroimmunology offers useful (although not unique) lessons about interdisciplinarity. Two types of interdisciplinarity are found in neuroimmunology: horizontal (integration of different disciplines – here neurobiology and immunology, to which one must actually add endocrinology and psychology) and vertical (integration of different levels of analysis, from molecules to systems, pertaining traditionally to distinct domains). Neuroimmunology also illustrates exemplarily the well-known fact that disciplines create mindsets and even norms of judgement and action. These mindsets are difficult to overcome in basic research as well as in the clinic. For example, few immunologists think about using neurostimulation to cure patients with autoimmune diseases, and until recently few psychiatrists considered using immune-based therapies to treat mental disorders. In addition to such epistemological questions, interdisciplinarity in neuroimmunology raises an ontological question: are there really three systems (nervous, immune, endocrine) in animals like us, or, as suggested, for example, by (Steinman 2012), are those only projections of our thought and language on the world? Personally, I see the epistemological issue as more stimulating and pressing than the ontological one, but both may converge to a large extent.

Second, what has been said here confirms the importance of adopting an extended view of immunity, as argued in the previous sections. In the context of neuroimmunology this extension has two dimensions. The first dimension has to do with the diversity of activities achieved by the immune system. To the various immunological activities already identified in the previous sections (development, clearance of debris, repair, and so on) one must add that the immune system could also influence the nervous system and cognitive functions. The second dimension is that the immune system is extended in another sense, namely insofar as it is connected with other systems (here the nervous and endocrine systems) and sometimes overlaps with them.

Third, our discussion suggests that immunology makes a significant contribution to the understanding of behavior and cognition and should therefore attract the attention of philosophers of neuroscience and philosophers of cognitive science. If the experimental results presented above are correct, then the immune system influences various feelings, behaviors, and cognitive processes either directly or indirectly (through its interactions with the CNS, PNS, endocrine system, and/or other systems). The immune system has an impact on sickness behavior (Konsman et al. 2002) and major depressive disorder (Dantzer 2018), and strategies are currently being developed to treat several behavioral disorders by targeting the immune system. Another major example is pain, a highly discussed topic in the philosophy of the neurosciences (Hardcastle 1997) and in general philosophy as well. Activated immune cells release pro-inflammatory cytokines, which sensitize sensory nerve endings, leading to an amplification and prolongation of pain; but the pain response also is downregulated via opioid-containing immune cells, as these cells release opioid peptides, which interact with opioid receptors on sensory nerves (Stein et al. 1990). All this corroborates the claim that pain is a highly complex process with several feedback loops and involving several bodily systems (Hardcastle 1997). Last but not least, many researchers propose that the immune system, via microglia, T cells, and cytokines, participates in cognition, particularly in learning and spatial memory.

All these data contribute to question brain-centered and more generally nervous system–centered views of behavior and cognition. Clearly, the nervous system remains crucial in all the processes described above, but it would nonetheless be inadequate to consider only the nervous system when trying to identify the biological basis of feelings, emotions, behaviors, and cognitive states. Understanding these processes requires an integrative approach in which the immune system could play an important role. Accordingly, neuroimmunology lends more weight to the idea of embodied cognition, that is, the idea that bodily elements distinct from the brain play a significant role in cognitive processing (Shapiro 2010). It also offers additional arguments to those who emphasize the importance of interoception (Craig 2002) insofar as it suggests that the brain can form a representation of the immunological status of peripheral tissues.

One may wonder, though, whether the immune system is an essential component of behavior and cognition. Here the answer will depend on what exactly is meant by this question. If the question is whether the immune system could influence behavior and cognition on its own, that is, independently of the (central and peripheral) nervous system, then the answer is probably negative. But if the question is whether the immune system is essential for proper functioning of the nervous system in some behavioral and cognitive activities, then the answer is affirmative, as exemplarily illustrated by microglia-mediated synaptic remodeling. More specifically, I suggest that the immune system can play three kinds of specific roles that are essential for the functioning of the nervous system and particularly for the realization of some cognitive processes.

First, the immune system is an *informant* for the nervous system. It provides vital information about infections, damage, and other perturbations that occur in any part of the organism. If, as proposed in Section 2, the capacity to respond to pathogens and other sources of damage constitutes one of the strongest evolutionary pressures on organisms, then conveying information to the nervous system about the immunological status of the host is vital. Second, the immune system is an *executant* for the nervous system: it realizes distinctive activities indispensable for the functioning of the nervous system, such as elimination of dead cells, repair, and so on. Third, and most crucially, the immune system is a *messenger* for the nervous system: not only does the nervous system resort to the molecular communication pathways of the immune system (cytokines) but it makes use of the unique feature of immune cells, namely their mobility, which allows them to reach any part of the organism and to deliver complex messages there. I suspect that, in coming years, cognitive functions mediated by this unique mobility of immune cells will be uncovered.

Current research in neuroimmunology raises other philosophically interesting issues, which could not be examined here due to space restrictions. These include how the nervous and the immune system interact in the construction of biological individuality and rethinking central physiological concepts such as homeostasis in light of the crosstalk between these two systems, among many other questions. Again, the aim of this section was simply to convince the reader that neuroimmunology is full of philosophical promise.

References

Ader R (2000) On the development of psychoneuroimmunology. *Eur J Pharmacol* 405:167–176

Ader R (1981) *Psychoneuroimmunology*. Academic Press, New York

Ader R, Cohen N (1985) High time for psychoimmunology. *Nature* 315:103–104. doi:10.1038/315103b0

Afik R, Zigmond E, Vugman M, et al. (2016) Tumor macrophages are pivotal constructors of tumor collagenous matrix. *Journal of Experimental Medicine* 213:2315–2331. doi:10.1084/jem.20151193

Aktipis CA, Boddy AM, Jansen G, et al. (2015) Cancer across the tree of life: cooperation and cheating in multicellularity. *Phil Trans R Soc B* 370:20140219. doi:10.1098/rstb.2014.0219

Anderson RM, May RM (1982) Coevolution of hosts and parasites. *Parasitology* 85 (Pt 2):411–426

Anderson W, Mackay IR (2014) Fashioning the immunological self: the biological individuality of F. Macfarlane Burnet. *J Hist Biol* 47:147–175. doi:10.1007/s10739-013-9352-1

Anderton SM, Wraith DC (2002) Selection and fine-tuning of the autoimmune T-cell repertoire. *Nat Rev Immunol* 2:487–498. doi:10.1038/nri842

Augustin R, Fraune S, Bosch TCG (2010) How Hydra senses and destroys microbes. *Semin Immunol* 22:54–58. doi:10.1016/j.smim.2009.11.002

Ayuso-Mateos JL, Vázquez-Barquero JL, Dowrick C, et al. (2001) Depressive disorders in Europe: prevalence figures from the ODIN study. *The British Journal of Psychiatry* 179:308–316. doi:10.1192/bjp.179.4.308

Bach J-F (2002) The effect of infections on susceptibility to autoimmune and allergic diseases. *N Engl J Med* 347:911–920. doi:10.1056/NEJMra020100

Bazin H (2011) Vaccination: a history from Lady Montagu to genetic engineering. *John Libbey Eurotext*, Montrouge

Belkaid Y, Harrison OJ (2017) Homeostatic immunity and the microbiota. *Immunity* 46:562–576. doi:10.1016/j.immuni.2017.04.008

Bertolaso M (2016) *Philosophy of Cancer: A Dynamic and Relational View*. Springer, Netherlands

Besedovsky H, Sorkin E (1977) Network of immune-neuroendocrine interactions. *Clin Exp Immunol* 27:1–12

Besedovsky HO, Rey AD (2007) Physiology of psychoneuroimmunology: a personal view. *Brain Behav Immun* 21:34–44. doi:10.1016/j.bbi.2006.09.008

Billingham RE, Brent L, Medawar PB (1953) Actively acquired tolerance of foreign cells. *Nature* 172:603–606. doi:10.1038/172603a0

Binnewies M, Roberts EW, Kersten K, et al. (2018) Understanding the tumor immune microenvironment (TIME) for effective therapy. *Nat Med* 24:541–550. doi:10.1038/s41591-018-0014-x

Bissell MJ, Hines WC (2011) Why don't we get more cancer? A proposed role of the microenvironment in restraining cancer progression. *Nat Med* 17:320–329. doi:10.1038/nm.2328

Bissell MJ, Radisky D (2001) Putting tumours in context. *Nat Rev Cancer* 1:46–54. doi:10.1038/35094059

Bosch TCG, McFall-Ngai MJ (2011) Metaorganisms as the new frontier. *Zoology (Jena)* 114:185–190. doi:10.1016/j.zool.2011.04.001

Bouchard F (2010) Symbiosis, lateral function transfer and the (many) saplings of life. *Biol Philos* 25:623–641. doi:10.1007/s10539-010-9209-3

Bretscher P, Cohn M (1970) A theory of self-nonself discrimination. *Science* 169:1042–1049

Bullmore E (2018) *The Inflamed Mind: A Radical New Approach to Depression*. Short Books, London

Burnet FM (1940) *Biological Aspects of Infectious Disease*. Macmillan, New York

Burnet FM (1969) *Cellular Immunology: Self and Notself*. Cambridge University Press, Cambridge

Burnet FM (1962) *The Integrity of the Body: A Discussion of Modern Immunological Ideas*. Harvard University Press, Cambridge, MA

Burnet FM (1960) Immunological recognition of self. *Nobel Lectures in Physiology or Medicine* 3:689–701

Burnet FM (1959) *The Clonal Selection Theory of Acquired Immunity*. Cambridge University Press, Cambridge

Burnet FM (1957) Cancer: a biological approach. *British Medical Journal* 1:1–7

Burnet FM (1970) *Immunological Surveillance*. Pergamon, Oxford

Burnet FM, Fenner F (1949) *The Production of Antibodies*, 2nd ed. Macmillan, Melbourne

Buss LW (1999) Slime molds, ascidians, and the utility of evolutionary theory. *Proc Natl Acad Sci U S A* 96:8801–8803

Buss LW (1987) *The Evolution of Individuality*. Princeton University Press, Princeton, NJ

Camacho-Arroyo I, López-Griego L, Morales-Montor J (2009) The role of cytokines in the regulation of neurotransmission. *Neuroimmunomodulation* 16:1–12. doi:10.1159/000179661

Cann SAH, Netten JP van, Netten C van (2003) Dr William Coley and tumour regression: a place in history or in the future. *Postgraduate Medical Journal* 79:672–680

Carosella ED, Pradeu T (2006) Transplantation and identity: a dangerous split? *Lancet* 368:183–184. doi:10.1016/S0140-6736(06)68938-1

Carson MJ, Doose JM, Melchior B, et al. (2006) CNS immune privilege: hiding in plain sight. *Immunol Rev* 213:48–65. doi:10.1111/j.1600-065X.2006.00441.x

Casadevall A, Pirofski L (1999) Host-pathogen interactions: redefining the basic concepts of virulence and pathogenicity. *Infect Immun* 67:3703–3713

Chavan SS, Pavlov VA, Tracey KJ (2017) Mechanisms and therapeutic relevance of neuro-immune communication. *Immunity* 46:927–942. doi:10.1016/j.immuni.2017.06.008

Chavan SS, Tracey KJ (2017) Essential neuroscience in immunology. *J Immunol* 198:3389–3397. doi:10.4049/jimmunol.1601613

Chen C, Itakura E, Nelson GM, et al. (2017) IL-17 is a neuromodulator of *Caenorhabditis elegans* sensory responses. *Nature* 542:43–48. doi:10.1038/nature20818

Chen DS, Mellman I (2017) Elements of cancer immunity and the cancer-immune set point. *Nature* 541:321–330. doi:10.1038/nature21349

Chen G, Zhuchenko O, Kuspa A (2007) Immune-like phagocyte activity in the social amoeba. *Science* 317:678–681. doi:10.1126/science.1143991

Chen YE, Fischbach MA, Belkaid Y (2018) Skin microbiota-host interactions. *Nature* 553:427–436. doi:10.1038/nature25177

Chiu L, Bazin T, Truchetet M-E, et al. (2017) Protective microbiota: from localized to long-reaching co-immunity. *Front Immunol* 8:. doi:10.3389/fimmu.2017.01678

Chow J, Tang H, Mazmanian SK (2011) Pathobionts of the gastrointestinal microbiota and inflammatory disease. *Current Opinion in Immunology* 23:473–480. doi:10.1016/j.coi.2011.07.010

Chu H, Mazmanian SK (2013) Innate immune recognition of the microbiota promotes host-microbial symbiosis. *Nat Immunol* 14:668–675. doi:10.1038/ni.2635

Clarke E (2011) The problem of biological individuality. *Biological Theory* 5:312–325

Cohen IR (2000) *Tending Adam's Garden: Evolving the Cognitive Immune Self.* Academic Press, San Diego

Cohen N (2006) The uses and abuses of psychoneuroimmunology: A global overview. *Brain, Behavior, and Immunity* 20:99–112. doi:10.1016/j.bbi.2005.09.008

Coley WB (1893) The treatment of malignant tumours by repeated inoculations of erysipelas with a report of ten original cases. *Am J Med Sci* 105:487–511

Craig AD (2002) How do you feel? Interoception: the sense of the physiological condition of the body. *Nature Reviews Neuroscience* 3:655–666. doi:10.1038/nrn894

Cremer S, Armitage SAO, Schmid-Hempel P (2007) Social immunity. *Curr Biol* 17:R693–702. doi:10.1016/j.cub.2007.06.008

Cummins R (1975) Functional Analysis. *The Journal of Philosophy* 72:741–765. doi:10.2307/2024640

da Costa LF (2001) Return of de-differentiation: why cancer is a developmental disease. *Curr Opin Oncol* 13:58–62

Dantzer R (2018) Neuroimmune interactions: From the brain to the immune system and vice versa. *Physiol Rev* 98:477–504. doi:10.1152/physrev.00039.2016

Dantzer R, O'Connor JC, Freund GG, et al. (2008) From inflammation to sickness and depression: when the immune system subjugates the brain. *Nat Rev Neurosci* 9:46–56. doi:10.1038/nrn2297

De Tomaso AW, Nyholm SV, Palmeri KJ, et al. (2005) Isolation and characterization of a protochordate histocompatibility locus. *Nature* 438:454–459. doi:10.1038/nature04150

de Visser KE, Eichten A, Coussens LM (2006) Paradoxical roles of the immune system during cancer development. *Nat Rev Cancer* 6:24–37. doi:10.1038/nrc1782

Dendrou CA, Fugger L, Friese MA (2015) Immunopathology of multiple sclerosis. *Nature Reviews Immunology* 15:545–558. doi:10.1038/nri3871

Derecki NC, Cardani AN, Yang CH, et al. (2010) Regulation of learning and memory by meningeal immunity: a key role for IL-4. *J Exp Med* 207:1067–1080. doi:10.1084/jem.20091419

Dolberg DS, Hollingsworth R, Hertle M, Bissell MJ (1985) Wounding and its role in RSV-mediated tumor formation. *Science* 230:676–678

Donaldson GP, Ladinsky MS, Yu KB, et al. (2018) Gut microbiota utilize immunoglobulin A for mucosal colonization. *Science* 360:795–800. doi:10.1126/science.aaq0926

Donaldson GP, Lee SM, Mazmanian SK (2016) Gut biogeography of the bacterial microbiota. *Nat Rev Micro* 14:20–32. doi:10.1038/nrmicro3552

Doron S, Melamed S, Ofir G, et al. (2018) Systematic discovery of antiphage defense systems in the microbial pangenome. *Science* eaar4120. doi:10.1126/science.aar4120

Dubernard J-M, Owen E, Herzberg G, et al. (1999) Human hand allograft: report on first 6 months. *The Lancet* 353:1315–1320. doi:10.1016/S0140-6736(99)02062-0

Dubernard JM, Owen ER, Lanzetta M, Hakim N (2001) What is happening with hand transplants? *The Lancet* 357:1711–1712. doi:10.1016/S0140-6736(00)04846-7

Dunn GP, Bruce AT, Ikeda H, et al. (2002) Cancer immunoediting: from immunosurveillance to tumor escape. *Nat Immunol* 3:991–998. doi:10.1038/ni1102-991

Dupré J (2010) The polygenomic organism. *The Sociological Review* 58:19–31. doi:10.1111/j.1467-954X.2010.01909.x

Dupré J, O'Malley M (2009) Varieties of living things: Life at the intersection of lineage and metabolism. *Philosophy & Theory in Biology* 1:. doi:http://dx.doi.org/10.3998/ptb.6959004.0001.003

Dustin ML, Colman DR (2002) Neural and immunological synaptic relations. *Science* 298:785–789. doi:10.1126/science.1076386

Dvorak HF (1986) Tumors: wounds that do not heal. Similarities between tumor stroma generation and wound healing. *N Engl J Med* 315:1650–1659. doi:10.1056/NEJM198612253152606

Dvorak HF (2015) Tumors: Wounds that do not heal – Redux. *Cancer Immunol Res* 3:1–11. doi:10.1158/2326-6066.CIR-14-0209

Eberl G (2010) A new vision of immunity: homeostasis of the superorganism. *Mucosal Immunology* 3:450–460. doi:10.1038/mi.2010.20

Eberl G (2016) Immunity by equilibrium. *Nat Rev Immunol* 16:524–532. doi:10.1038/nri.2016.75

Egeblad M, Nakasone ES, Werb Z (2010) Tumors as organs: complex tissues that interface with the entire organism. *Dev Cell* 18:884–901. doi:10.1016/j.devcel.2010.05.012

Ehrlich P (1909) Ueber den jetzigen Stand der Karzinomforschung. *Nederlands Tijdschrift voor Geneeskunde* 5:273–290

Eming SA, Martin P, Tomic-Canic M (2014) Wound repair and regeneration: mechanisms, signaling, and translation. *Sci Transl Med* 6:265sr6. doi:10.1126/scitranslmed.3009337

Eom DS, Parichy DM (2017) A macrophage relay for long-distance signaling during postembryonic tissue remodeling. *Science* 355:1317–1320. doi:10.1126/science.aal2745

European Directorate for the Quality of Medicines & HealthCare (2015) Newsletter Transplant: International figures on donation and transplantation https://www.edqm.eu/sites/default/files/newsletter_transplant_volume_21_september_2016.pdf

Fernández-Sánchez ME, Barbier S, Whitehead J, et al. (2015) Mechanical induction of the tumorigenic β-catenin pathway by tumour growth pressure. *Nature* 523:92–95. doi:10.1038/nature14329

Filiano AJ, Xu Y, Tustison NJ, et al. (2016) Unexpected role of interferon-γ in regulating neuronal connectivity and social behaviour. *Nature* 535:425–429. doi:10.1038/nature18626

Finlay BB, McFadden G (2006) Anti-immunology: Evasion of the host immune system by bacterial and viral pathogens. *Cell* 124:767–782. doi:10.1016/j.cell.2006.01.034

Fisher R, Pusztai L, Swanton C (2013) Cancer heterogeneity: implications for targeted therapeutics. *British Journal of Cancer* 108:479–485. doi:10.1038/bjc.2012.581

Flajnik MF, Du Pasquier L (2004) Evolution of innate and adaptive immunity: can we draw a line? *Trends in Immunology* 25:640–644

Folkman J, Kalluri R (2004) Cancer without disease. *Nature* 427:787. doi:10.1038/427787a

Frank SA (2007) *Dynamics of Cancer: Incidence, Inheritance, and Evolution*. Princeton University Press, Princeton, NJ

Fridlender ZG, Sun J, Kim S, et al. (2009) Polarization of tumor-associated neutrophil phenotype by TGF-beta: "N1" versus "N2" TAN. *Cancer Cell* 16:183–194. doi:10.1016/j.ccr.2009.06.017

Fulton RB, Hamilton SE, Xing Y, et al. (2015) The TCR's sensitivity to self peptide-MHC dictates the ability of naive CD8+ T cells to respond to foreign antigens. *Nat Immunol* 16:107–117. doi:10.1038/ni.3043

Fung TC, Olson CA, Hsiao EY (2017) Interactions between the microbiota, immune and nervous systems in health and disease. *Nat Neurosci* 20:145–155. doi:10.1038/nn.4476

Galluzzi L, Senovilla L, Zitvogel L, Kroemer G (2012) The secret ally: immunostimulation by anticancer drugs. *Nat Rev Drug Discov* 11:215–233. doi:10.1038/nrd3626

Galon J, Costes A, Sanchez-Cabo F, et al. (2006) Type, Density, and Location of Immune Cells Within Human Colorectal Tumors Predict Clinical Outcome. *Science* 313:1960–1964. doi:10.1126/science.1129139

Germain P-L (2012) Cancer cells and adaptive explanations. *Biol Philos* 27:785–810. doi:10.1007/s10539-012-9334-2

Gilbert SF (2002) The genome in its ecological context: philosophical perspectives on interspecies epigenesis. *Ann N Y Acad Sci* 981:202–218

Gilbert SF, Sapp J, Tauber AI (2012) A symbiotic view of life: we have never been individuals. *Q Rev Biol* 87:325–341

Gill SR, Pop M, DeBoy RT, et al. (2006) Metagenomic analysis of the human distal gut microbiome. *Science* 312:1355–1359. doi:10.1126/science.1124234

Godfrey-Smith P (2009) *Darwinian Populations and Natural Selection*. Oxford University Press, Oxford

Godfrey-Smith P (2016) *Other Minds: The Octopus, the Sea, and the Deep Origins of Consciousness*, First edition. Farrar, Straus and Giroux, New York

Goodnow CC, Sprent J, de St Groth BF, Vinuesa CG (2005) Cellular and genetic mechanisms of self tolerance and autoimmunity. *Nature* 435:590–597. doi:10.1038/nature03724

Gordon S (2003) Alternative activation of macrophages. *Nat Rev Immunol* 3:23–35. doi:10.1038/nri978

Goubau D, Deddouche S, Reis e Sousa C (2013) Cytosolic sensing of viruses. *Immunity* 38:855–869. doi:10.1016/j.immuni.2013.05.007

Greaves M (2007) Darwinian medicine: a case for cancer. *Nat Rev Cancer* 7:213–221. doi:10.1038/nrc2071

Greaves M, Maley CC (2012) Clonal evolution in cancer. *Nature* 481:306–313. doi:10.1038/nature10762

Grossman Z, Paul WE (1992) Adaptive cellular interactions in the immune system: the tunable activation threshold and the significance of subthreshold responses. *Proceedings of the National Academy of Sciences* 89:10365–10369

Guay A, Pradeu T (2016) *Individuals Across the Sciences*. Oxford University Press, New York

Gurtner GC, Werner S, Barrandon Y, Longaker MT (2008) Wound repair and regeneration. *Nature* 453:314–321. doi:10.1038/nature07039

Haber MH (2016) The individuality thesis (3 ways). *Biol Philos* 31:913–930. doi:10.1007/s10539-016-9548-9

Hanahan D, Weinberg RA (2011) Hallmarks of cancer: the next generation. *Cell* 144:646–674. doi:10.1016/j.cell.2011.02.013

Hanahan D, Weinberg RA (2000) The hallmarks of cancer. *Cell* 100:57–70. doi:10.1016/S0092-8674(00)81683-9

Hansson GK, Hermansson A (2011) The immune system in atherosclerosis. *Nature* Immunology 12:204–212. doi:10.1038/ni.2001

Hardcastle VG (1997) When a pain is not. *Journal of Philosophy* 94:381–409

Hartung HP, Toyka KV (1983) Activation of macrophages by substance P: induction of oxidative burst and thromboxane release. *Eur J Pharmacol* 89:301–305

Haslam A, Prasad V (2019) Estimation of the percentage of US patients with cancer who are eligible for and respond to checkpoint inhibitor immunotherapy drugs. *JAMA Netw Open* 2:e192535–e192535. doi:10.1001/jamanetworkopen.2019.2535

Heneka MT, Kummer MP, Latz E (2014) Innate immune activation in neurodegenerative disease. *Nature Reviews Immunology* 14:463–477. doi:10.1038/nri3705

Heppner FL, Ransohoff RM, Becher B (2015) Immune attack: the role of inflammation in Alzheimer disease. *Nature Reviews Neuroscience* 16:358–372. doi:10.1038/nrn3880

Hille F, Richter H, Wong SP, et al. (2018) The biology of CRISPR-Cas: Backward and forward. *Cell* 172:1239–1259. doi:10.1016/j.cell.2017.11.032

Hodi FS, O'Day SJ, McDermott DF, et al. (2010) Improved survival with Ipilimumab in patients with metastatic melanoma. *New England Journal of Medicine* 363:711–723. doi:10.1056/NEJMoa1003466

Hoeijmakers JHJ (2001) Genome maintenance mechanisms for preventing cancer. In: *Nature*. www.nature.com/articles/35077232. Accessed October 9, 2018

Hogquist KA, Jameson SC (2014) The self-obsession of T cells: how TCR signaling thresholds affect fate "decisions" and effector function. *Nature Immunology* 15:815–823. doi:10.1038/ni.2938

Hooper LV, Gordon JI (2001) Commensal host-bacterial relationships in the gut. *Science* 292:1115–1118

Horvath P, Barrangou R (2010) CRISPR/Cas, the immune system of bacteria and Archaea. *Science* 327:167–170. doi:10.1126/science.1179555

Houghton AN (1994) Cancer antigens: immune recognition of self and altered self. *The Journal of Experimental Medicine* 180:1–4. doi:10.1084/jem.180.1.1

Huang S, Ernberg I, Kauffman S (2009) Cancer attractors: A systems view of tumors from a gene network dynamics and developmental perspective. *Seminars in Cell & Developmental Biology* 20:869–876. doi:10.1016/j.semcdb.2009.07.003

Hull D (1992) Individual. In: Keller EF, Lloyd EA (eds.) *Keywords in Evolutionary Biology*. Harvard University Press, Cambridge, MA, pp. 181–187

Hull D (1980) Individuality and selection. *Annual Review of Ecology and Systematics* 11:311–332. doi:10.1146/annurev.es.11.110180.001523

Hull DL (1978) A matter of individuality. *Philosophy of Science* 45:335–360

Huneman P (2014) Individuality as a theoretical scheme. II. About the weak individuality of organisms and ecosystems. *Biol Theory* 9:374–381. doi:10.1007/s13752-014-0193-8

Jackson SA, McKenzie RE, Fagerlund RD, et al. (2017) CRISPR-Cas: Adapting to change. *Science* 356:eaal5056. doi:10.1126/science.aal5056

Jain RK (2005) Normalization of tumor vasculature: an emerging concept in antiangiogenic therapy. *Science* 307:58–62. doi:10.1126/science.1104819

Janeway CA (1989) Approaching the asymptote? Evolution and revolution in immunology. *Cold Spring Harb Symp Quant Biol*:1–13

Janeway CA (2001) How the immune system protects the host from infection. *Microbes Infect* 3:1167–1171

Janeway CA (1992) The immune system evolved to discriminate infectious nonself from noninfectious self. *Immunology Today* 13:11–16. doi:10.1016/0167-5699(92)90198-G

Jerne NK (1985) The generative grammar of the immune system. *EMBO J* 4:847–852

Jiang H, Chess L (2009) How the immune system achieves self-nonself discrimination during adaptive immunity. *Adv Immunol* 102:95–133. doi:10.1016/S0065-2776(09)01202-4

Jones B, Shipley E, Arnold KE (2018) Social immunity in honeybees – Density dependence, diet, and body mass trade-offs. *Ecol Evol* 8:4852–4859. doi:10.1002/ece3.4011

Jones JDG, Dangl JL (2006) The plant immune system. *Nature* 444:323–329. doi:10.1038/nature05286

Joyce JA (2005) Therapeutic targeting of the tumor microenvironment. *Cancer Cell* 7:513–520. doi:10.1016/j.ccr.2005.05.024

Joyce JA, Fearon DT (2015) T cell exclusion, immune privilege, and the tumor microenvironment. *Science* 348:74–80. doi:10.1126/science.aaa6204

June CH, Sadelain M (2018) Chimeric antigen receptor therapy. *N Engl J Med* 379:64–73. doi:10.1056/NEJMra1706169

Kahan BD (2003) Individuality: the barrier to optimal immunosuppression. *Nature Reviews Immunology* 3:831–838. doi:10.1038/nri1204

Keijzer F, Duijn M van, Lyon P (2013) What nervous systems do: early evolution, input–output, and the skin brain thesis. *Adaptive Behavior* 21:67–85. doi:10.1177/1059712312465330

Kelly PN (2018) The cancer immunotherapy revolution. *Science* 359:1344–1345. doi:10.1126/science.359.6382.1344

Khandaker GM, Cousins L, Deakin J, et al. (2015) Inflammation and immunity in schizophrenia: implications for pathophysiology and treatment. *Lancet Psychiatry* 2:258–270. doi:10.1016/S2215-0366(14)00122-9

Kiecolt-Glaser JK, McGuire L, Robles TF, Glaser R (2002) Psychoneuroimmunology and psychosomatic medicine: back to the future. *Psychosom Med* 64:15–28

Kiers ET, West SA (2015) Evolving new organisms via symbiosis. *Science* 348:392–394. doi:10.1126/science.aaa9605

Kioussis D, Pachnis V (2009) Immune and nervous systems: more than just a superficial similarity? *Immunity* 31:705–710. doi:10.1016/j.immuni.2009.09.009

Kipnis J (2016) Multifaceted interactions between adaptive immunity and the central nervous system. *Science* 353:766–771. doi:10.1126/science.aag2638

Kipnis J, Cohen H, Cardon M, et al. (2004) T cell deficiency leads to cognitive dysfunction: Implications for therapeutic vaccination for schizophrenia and other psychiatric conditions. *Proc Natl Acad Sci U S A* 101:8180–8185. doi:10.1073/pnas.0402268101

Kipnis J, Gadani S, Derecki NC (2012) Pro-cognitive properties of T cells. *Nat Rev Immunol* 12:663–669. doi:10.1038/nri3280

Kitano H (2004a) Cancer as a robust system: implications for anticancer therapy. *Nat Rev Cancer* 4:227–235. doi:10.1038/nrc1300

Kitano H (2004b) Biological robustness. *Nat Rev Genet* 5:826–837. doi:10.1038/nrg1471

Klein G, Imreh S, Zabarovsky ER (2007) Why Do We Not All Die of Cancer at an Early Age? In: *Advances in Cancer Research*. Academic Press, pp. 1–16

Klein J (1982) *Immunology: The Science of Self-Nonself Discrimination*. Wiley, New York

Koebel CM, Vermi W, Swann JB, et al. (2007) Adaptive immunity maintains occult cancer in an equilibrium state. *Nature* 450:903–907. doi:10.1038/nature06309

Konsman JP, Parnet P, Dantzer R (2002) Cytokine-induced sickness behaviour: mechanisms and implications. *Trends in Neurosciences* 25:154–159. doi:10.1016/S0166-2236(00)02088-9

Koonin EV (2019) CRISPR: a new principle of genome engineering linked to conceptual shifts in evolutionary biology. *Biol Philos* 34:9. doi:10.1007/s10539-018-9658-7

Koopman FA, Chavan SS, Miljko S, et al. (2016) Vagus nerve stimulation inhibits cytokine production and attenuates disease severity in rheumatoid arthritis. PNAS 113:8284–8289. doi:10.1073/pnas.1605635113

Kovaka K (2015) Biological individuality and scientific practice. *Philosophy of Science* 82:1092–1103. doi:10.1086/683443

Kumar V, Patel S, Tcyganov E, Gabrilovich DI (2016) The nature of myeloid-derived suppressor cells in the tumor microenvironment. *Trends in Immunology* 37:208–220. doi:10.1016/j.it.2016.01.004

Langman RE, Cohn M (2000) Editorial introduction. *Seminars in Immunology* 12:159–162. doi:10.1006/smim.2000.0227

Lanier LL, Sun JC (2009) Do the terms innate and adaptive immunity create conceptual barriers? *Nat Rev Immunol* 9:302–303. doi:10.1038/nri2547

Laplane L (2016) *Cancer Stem Cells: Philosophy and Therapies*. Harvard University Press, Cambridge, MA

Laplane L, Duluc D, Larmonier N, et al. (2018) The multiple layers of the tumor environment. *Trends in Cancer*. doi:10.1016/j.trecan.2018.10.002

Laplane L, Mantovani P, Adolphs R, Chang H, Mantovani A, McFall-Ngai M, Rovelli C, Sober E, Pradeu T. 2019. Why science needs philosophy. *PNAS* 116:3948–3952. doi:10.1073/pnas.1900357116

Laurent P, Jolivel V, Manicki P, et al. (2017) Immune-mediated repair: a matter of plasticity. *Front Immunol* 8:. doi:10.3389/fimmu.2017.00454

Leach DR, Krummel MF, Allison JP (1996) Enhancement of antitumor immunity by CTLA-4 blockade. *Science* 271:1734–1736

Lean C, Plutynski A (2016) The evolution of failure: explaining cancer as an evolutionary process. *Biol Philos* 31:39–57. doi:10.1007/s10539-015-9511-1

Leboyer M, Berk M, Yolken RH, et al. (2016) Immuno-psychiatry: an agenda for clinical practice and innovative research. *BMC Medicine* 14:173. doi:10.1186/s12916-016-0712-5

Lee H, Brott BK, Kirkby LA, et al. (2014) Synapse elimination and learning rules co-regulated by MHC class I H2-D^b. *Nature* 509:195–200. doi:10.1038/nature13154

Lehuen A, Diana J, Zaccone P, Cooke A (2010) Immune cell crosstalk in type 1 diabetes. *Nature Reviews Immunology* 10:501–513. doi:10.1038/nri2787

Lemaitre B, Hoffmann J (2007) The host defense of *Drosophila melanogaster*. *Annu Rev Immunol* 25:697–743. doi:10.1146/annurev.immunol.25.022106.141615

Lesokhin AM, Callahan MK, Postow MA, Wolchok JD (2015) On being less tolerant: Enhanced cancer immunosurveillance enabled by targeting checkpoints and agonists of T cell activation. *Science Translational Medicine* 7:280sr1-280sr1. doi:10.1126/scitranslmed.3010274

Letai A (2017) Apoptosis and Cancer. *Annual Review of Cancer Biology* 1:275–294. doi:10.1146/annurev-cancerbio-050216-121933

Liblau RS (2018) Put to sleep by immune cells. *Nature*. doi:10.1038/d41586-018-06666-w

Lidgard S, Nyhart LK (eds.) (2017) *Biological Individuality: Integrating Scientific, Philosophical, and Historical Perspectives*. The University of Chicago Press, Chicago

Loeb L (1937) The biological basis of individuality. *Science* 86:1–5

Louveau A, Harris TH, Kipnis J (2015a) Revisiting the mechanisms of CNS immune privilege. *Trends in Immunology* 36:569–577. doi:10.1016/j.it.2015.08.006

Louveau A, Smirnov I, Keyes TJ, et al. (2015b) Structural and functional features of central nervous system lymphatic vessels. *Nature* 523:337–341. doi:10.1038/nature14432

Löwy I (1991) The Immunological Construction of the Self. In: Tauber AI (ed.) *Organism and the Origins of Self.* Kluwer, Dordrecht, pp. 3–75

Lumeng CN, Saltiel AR (2011) Inflammatory links between obesity and metabolic disease. *J Clin Invest* 121:2111–2117. doi:10.1172/JCI57132

Maddox J (1984) Psychoimmunology before its time. *Nature* 309:400. doi:10.1038/309400a0

Maman S, Witz IP (2018) A history of exploring cancer in context. *Nature Reviews Cancer* 18:359–376. doi:10.1038/s41568-018-0006-7

Mantovani A, Allavena P, Sica A, Balkwill F (2008) Cancer-related inflammation. *Nature* 454:436–444. doi:10.1038/nature07205

Mantovani A, Biswas SK, Galdiero MR, et al. (2013) Macrophage plasticity and polarization in tissue repair and remodelling. *J Pathol* 229:176–185. doi:10.1002/path.4133

Mantovani A, Bottazzi B, Colotta F, et al. (1992) The origin and function of tumor-associated macrophages. *Immunol Today* 13:265–270. doi:10.1016/0167-5699(92)90008-U

Matthen M, Levy E (1984) Teleology, error, and the human immune system. *The Journal of Philosophy* 81:351–372. doi:10.2307/2026291

Matzinger P (2002) The danger model: A renewed sense of self. *Science* 296:301–305

Maynard Smith J, Szathmáry E (1995) *The Major Transitions in Evolution.* WHFreeman Spektrum, Oxford; New York

McFall-Ngai M, Hadfield MG, Bosch TCG, et al. (2013) Animals in a bacterial world, a new imperative for the life sciences. *Proc Natl Acad Sci USA* 110:3229–3236. doi:10.1073/pnas.1218525110

McFall-Ngai MJ (2002) Unseen forces: the influence of bacteria on animal development. *Developmental Biology* 242:1–14. doi:10.1006/dbio.2001.0522

McFarland H, Reich DS, Jacobson S (2017) Introduction to the Special Issue "40 Years-Neuroimmunology." *Journal of Neuroimmunology* 304:1. doi:10.1016/j.jneuroim.2017.01.022

Medawar PB (1957) *The Uniqueness of the Individual.* Methuen, Londres

Melander P (1993) How not to explain the errors of the immune system. *Philosophy of Science* 60:223–241. doi:10.1086/289730

Mellor AL, Munn DH (2008) Creating immune privilege: active local suppression that benefits friends, but protects foes. *Nat Rev Immunol* 8:74–80. doi:10.1038/nri2233

Melo FDSE, Vermeulen L, Fessler E, Medema JP (2013) Cancer heterogeneity—a multifaceted view. *EMBO Reports* 14:686–695. doi:10.1038/embor.2013.92

Méthot P-O, Alizon S (2014) What is a pathogen? Toward a process view of host-parasite interactions. *Virulence* 5:775–785. doi:10.4161/21505594.2014.960726

Michell-Robinson MA, Touil H, Healy LM, et al. (2015) Roles of microglia in brain development, tissue maintenance and repair. *Brain* 138:1138–1159. doi:10.1093/brain/awv066

Michod RE (1999) *Darwinian Dynamics: Evolutionary Transitions in Fitness and Individuality.* Princeton University Press, Princeton, NJ

Miller AH, Raison CL (2016) The role of inflammation in depression: from evolutionary imperative to modern treatment target. *Nat Rev Immunol* 16:22–34. doi:10.1038/nri.2015.5

Miller BJ, Buckley PF (2017) Monoclonal antibody immunotherapy in psychiatric disorders. *The Lancet Psychiatry* 4:13–15. doi:10.1016/S2215-0366(16)30366-2

Miller G (2009) On the origin of the nervous system. *Science* 325:24–26. doi:10.1126/science.325_24

Moreau J-F, Pradeu T, GRignolio A, et al. The emerging role of ECM crosslinking in T cell mobility as a hallmark of immunosenescence in humans. *Ageing Research Reviews*. doi:10.1016/j.arr.2016.11.005

Moulin A-M (1991) Le Dernier langage de la médecine. PUF, Paris

Mueller KL, Hines PJ, Travis J (2016) Neuroimmunology. *Science* 353:760–761. doi:10.1126/science.353.6301.760

Müller WEG (2003) The origin of metazoan complexity: porifera as integrated animals. *Integr Comp Biol* 43:3–10. doi:10.1093/icb/43.1.3

Muraille E (2013) Redefining the immune system as a social interface for cooperative processes. *PLoS Pathog* 9:e1003203. doi:10.1371/journal.ppat.1003203

Murdoch C, Muthana M, Coffelt SB, Lewis CE (2008) The role of myeloid cells in the promotion of tumour angiogenesis. *Nat Rev Cancer* 8:618–631. doi:10.1038/nrc2444

Nagata S (2018) Apoptosis and clearance of apoptotic cells. *Annual Review of Immunology* 36:489–517. doi:10.1146/annurev-immunol-042617-053010

Nimmerjahn A, Kirchhoff F, Helmchen F (2005) Resting microglial cells are highly dynamic surveillants of brain parenchyma in vivo. *Science* 308:1314–1318. doi:10.1126/science.1110647

Nowell PC (1976) The clonal evolution of tumor cell populations. *Science* 194:23–28. doi:10.1126/science.959840

Nuñez JK, Harrington LB, Kranzusch PJ, et al. (2015) Foreign DNA capture during CRISPR–Cas adaptive immunity. *Nature* 527:535–538. doi:10.1038/nature15760

Okabe Y, Medzhitov R (2016) Tissue biology perspective on macrophages. *Nat Immunol* 17:9–17. doi:10.1038/ni.3320

Okasha S (2006) *Evolution and the Levels of Selection*. Clarendon Press; Oxford University Press, Oxford; NY

Ordovas-Montanes J, Rakoff-Nahoum S, Huang S, et al. (2015) The regulation of immunological processes by peripheral neurons in homeostasis and disease. *Trends Immunol* 36:578–604. doi:10.1016/j.it.2015.08.007

Osimo EF, Cardinal RN, Jones PB, Khandaker GM (2018) Prevalence and correlates of low-grade systemic inflammation in adult psychiatric inpatients: An electronic health record-based study. *Psychoneuroendocrinology* 91:226–234. doi:10.1016/j.psyneuen.2018.02.031

Ottaviani E, Malagoli D, Franceschi C (2007) Common evolutionary origin of the immune and neuroendocrine systems: from morphological and functional evidence to in silico approaches. *Trends in Immunology* 28:497–502. doi:10.1016/j.it.2007.08.007

Owen RD (1945) Immunogenetic consequences of vascular anastomoses between bovine twins. *Science* 102:400–401. doi:10.1126/science.102.2651.400

Pamer EG (2014) Fecal microbiota transplantation: effectiveness, complexities, and lingering concerns. *Mucosal Immunology* 7:210–214. doi:10.1038/mi.2013.117

Pamer EG (2016) Resurrecting the intestinal microbiota to combat antibiotic-resistant pathogens. *Science* 352:535–538. doi:10.1126/science.aad9382

Pancer Z, Cooper MD (2006) The evolution of adaptive immunity. *Annu Rev Immunol* 24:497–518. doi:10.1146/annurev.immunol.24.021605.090542

Pariante CM (2015) Psychoneuroimmunology or immunopsychiatry? *Lancet Psychiatry* 2:197–199. doi:10.1016/S2215-0366(15)00042-5

Parkhurst CN, Yang G, Ninan I, et al. (2013) Microglia promote learning-dependent synapse formation through brain-derived neurotrophic factor. *Cell* 155:1596–1609. doi:10.1016/j.cell.2013.11.030

Pastor-Pareja JC, Wu M, Xu T (2008) An innate immune response of blood cells to tumors and tissue damage in Drosophila. *Disease Models & Mechanisms* dmm.000950. doi:10.1242/dmm.000950

Pauken KE, Wherry EJ (2015) Overcoming T cell exhaustion in infection and cancer. *Trends in Immunology* 36:265–276. doi:10.1016/j.it.2015.02.008

Paul WE (2015) *Immunity*. Johns Hopkins University Press, Baltimore

Petrovsky N (2001) Towards a unified model of neuroendocrine–immune interaction. *Immunol Cell Biol* 79:350–357. doi:10.1046/j.1440-1711.2001.01029.x

Plutynski A (2018) *Explaining Cancer: Finding Order in Disorder.* Oxford University Press, Oxford, New York

Postow MA, Sidlow R, Hellmann MD (2018) Immune-related adverse events associated with immune checkpoint blockade. *New England Journal of Medicine* 378:158–168. doi:10.1056/NEJMra1703481

Pradeu T (2012) *The Limits of the Self: Immunology and Biological Identity.* Oxford University Press, New York

Pradeu T (2013) Immunity and the emergence of individuality. In: Bouchard F, Huneman P (eds.) *From Groups to Individuals: Evolution and Emerging Individuality.* MIT Press, Cambridge, MA, pp. 77–96

Pradeu T (2016a) The many faces of biological individuality. *Biol Philos* 31:761–773. doi:10.1007/s10539-016-9553-z

Pradeu T (2010) What is an organism? An immunological answer. *History and Philosophy of the Life Sciences* 32:247–268

Pradeu T (2016b) Organisms or biological individuals? Combining physiological and evolutionary individuality. *Biol Philos* 31:797–817. doi:10.1007/s10539-016-9551-1

Pradeu T, Carosella E (2006a) On the definition of a criterion of immunogenicity. *Proceedings of the National Academy of Sciences USA* 103:17858–17861

Pradeu T, Carosella E (2006b) The self model and the conception of biological identity in immunology. *Biology and Philosophy* 21:235–252

Pradeu T, Du Pasquier L (2018) Immunological memory: What's in a name? *Immunol Rev* 283:7–20. doi:10.1111/imr.12652

Pradeu T, Jaeger S, Vivier E (2013) The speed of change: towards a discontinuity theory of immunity? *Nat Rev Immunol* 13:764–769. doi:10.1038/nri3521

Prendergast GC (2012) Immunological thought in the mainstream of cancer research: Past divorce, recent remarriage and elective affinities of the future. *Oncoimmunology* 1:793–797. doi:10.4161/onci.20909

Qian B-Z, Pollard JW (2010) Macrophage diversity enhances tumor progression and metastasis. *Cell* 141:39–51. doi:10.1016/j.cell.2010.03.014

Queller DC, Strassmann JE (2009) Beyond society: the evolution of organismality. *Philos Trans R Soc Lond, B, Biol Sci* 364:3143–3155. doi:10.1098/rstb.2009.0095

Radisky D, Hagios C, Bissell MJ (2001) Tumors are unique organs defined by abnormal signaling and context. *Semin Cancer Biol* 11:87–95. doi:10.1006/scbi.2000.0360

Raison CL, Capuron L, Miller AH (2006) Cytokines sing the blues: inflammation and the pathogenesis of depression. *Trends Immunol* 27:24–31. doi:10.1016/j.it.2005.11.006

Rankin LC, Artis D (2018) Beyond host defense: emerging functions of the immune system in regulating complex tissue physiology. *Cell* 173:554–567. doi:10.1016/j.cell.2018.03.013

Ransohoff RM (2009) Chemokines and chemokine receptors: standing at the crossroads of immunobiology and neurobiology. *Immunity* 31:711–721. doi:10.1016/j.immuni.2009.09.010

Ribas A, Wolchok JD (2018) Cancer immunotherapy using checkpoint blockade. *Science* 359:1350–1355. doi:10.1126/science.aar4060

Richet CR (1894) *La défense de l'organisme: cours de physiologie de la Faculté de médecine (1893–1894)*. Typographie Chamerot et Renouard, Paris

Ricklin D, Hajishengallis G, Yang K, Lambris JD (2010) Complement: a key system for immune surveillance and homeostasis. *Nat Immunol* 11:785–797. doi:10.1038/ni.1923

Robert J (2010) Comparative study of tumorigenesis and tumor immunity in invertebrates and nonmammalian vertebrates. *Developmental & Comparative Immunology* 34:915–925. doi:10.1016/j.dci.2010.05.011

Round JL, Mazmanian SK (2009) The gut microbiota shapes intestinal immune responses during health and disease. *Nat Rev Immunol* 9:313–323. doi:10.1038/nri2515

Salter MW, Beggs S (2014) Sublime microglia: expanding roles for the guardians of the CNS. *Cell* 158:15–24. doi:10.1016/j.cell.2014.06.008

Sanmamed MF, Chen L (2018) A paradigm shift in cancer immunotherapy: from enhancement to normalization. *Cell* 175:313–326. doi:10.1016/j.cell.2018.09.035

Sansonetti PJ, Di Santo JP (2007) Debugging how bacteria manipulate the immune response. *Immunity* 26:149–161. doi:10.1016/j.immuni.2007.02.004

Sansonetti PJ, Medzhitov R (2009) Learning tolerance while fighting ignorance. *Cell* 138:416–420. doi:10.1016/j.cell.2009.07.024

Schäfer M, Werner S (2008) Cancer as an overhealing wound: an old hypothesis revisited. *Nat Rev Mol Cell Biol* 9:628–638. doi:10.1038/nrm2455

Schaffner KF (1992) Theory change in immunology. Part II: The clonal selection theory. *Theor Med* 13:191–216

Schou M, Sæther SG, Borowski K, et al. (2016) Prevalence of serum anti-neuronal autoantibodies in patients admitted to acute psychiatric care. *Psychol Med* 46:3303–3313. doi:10.1017/S0033291716002038

Schreiber RD, Old LJ, Smyth MJ (2011) Cancer immunoediting: integrating immunity's roles in cancer suppression and promotion. *Science* 331:1565–1570. doi:10.1126/science.1203486

Schretter CE, Vielmetter J, Bartos I, et al. (2018a) A gut microbial factor modulates locomotor behaviour in *Drosophila*. *Nature*. doi:10.1038/s41586-018-0634-9

Schretter CE, Vielmetter J, Bartos I, et al. (2018b) A gut microbial factor modulates locomotor behaviour in *Drosophila*. *Nature* 563:402. doi:10.1038/s41586-018-0634-9

Schulenburg H, Kurtz J, Moret Y, Siva-Jothy MT (2009) Introduction. Ecological immunology. *Philos Trans R Soc Lond B Biol Sci* 364:3–14. doi:10.1098/rstb.2008.0249

Schumacher TN, Schreiber RD (2015) Neoantigens in cancer immunotherapy. *Science* 348:69–74. doi:10.1126/science.aaa4971

Sender R, Fuchs S, Milo R (2016) Are we really vastly outnumbered? Revisiting the ratio of bacterial to host cells in humans. *Cell* 164:337–340. doi:10.1016/j.cell.2016.01.013

Shah J, Zeier J (2013) Long-distance communication and signal amplification in systemic acquired resistance. *Front Plant Sci* 4:. doi:10.3389/fpls.2013.00030

Shankaran V, Ikeda H, Bruce AT, et al. (2001) IFNgamma and lymphocytes prevent primary tumour development and shape tumour immunogenicity. *Nature* 410:1107–1111. doi:10.1038/35074122

Shapiro LA (2010) *Embodied Cognition*. Routledge, New York

Sharon G, Sampson TR, Geschwind DH, Mazmanian SK (2016) The central nervous system and the gut microbiome. *Cell* 167:915–932. doi:10.1016/j.cell.2016.10.027

Sherr CJ (1996) Cancer cell cycles. *Science* 274:1672–1677

Siemionow M (2016) The miracle of face transplantation after 10 years. *Br Med Bull* 120:5–14. doi:10.1093/bmb/ldw045

Silverstein AM (2009) *A History of Immunology*, 2nd ed. Academic Press, London ; Burlington, MA

Silverstein AM (2002) The Clonal Selection Theory: what it really is and why modern challenges are misplaced. *Nat Immunol* 3:793–796. doi:10.1038/ni0902-793

Simon HA (Herbert A (1969) *The Sciences of the Artificial*. MIT Press, Cambridge

Sivik T, Byrne D, Christodoulou CN, et al. (eds.) (2002) *Psycho-neuro-endocrino-immunology (PNEI): a common language for the whole human body: proceedings of the 16th World Congress on Psychosomatic Medicine, held in Göteborg, Sweden, 24th–29th August 2001*. Elsevier, Amsterdam; Boston

Sober E (1991) Organisms, Individuals, and Units of Selection. In: Tauber AI (ed.) *Organism and the Origins of Self*. Springer Netherlands, Dordrecht, pp. 275–296

Sparkman NL, Buchanan JB, Heyen JRR, et al. (2006) Interleukin-6 facilitates lipopolysaccharide-induced disruption in working memory and expression of other proinflammatory cytokines in hippocampal neuronal cell layers. *J Neurosci* 26:10709–10716. doi:10.1523/JNEUROSCI.3376-06.2006

Spoel SH, Dong X (2012) How do plants achieve immunity? Defence without specialized immune cells. *Nat Rev Immunol* 12:89–100. doi:10.1038/nri3141

Stearns SC, Koella JC (2008) *Evolution in Health and Disease*, 2nd ed. Oxford University Press, Oxford; New York

Stefanová I, Dorfman JR, Germain RN (2002) Self-recognition promotes the foreign antigen sensitivity of naive T lymphocytes. *Nature* 420:429–434. doi:10.1038/nature01146

Stein C, Hassan AH, Przewłocki R, et al. (1990) Opioids from immunocytes interact with receptors on sensory nerves to inhibit nociception in inflammation. *PNAS* 87:5935–5939. doi:10.1073/pnas.87.15.5935

Steinman L (2004) Elaborate interactions between the immune and nervous systems. *Nat Immunol* 5:575–581. doi:10.1038/ni1078

Steinman L (2012) Lessons learned at the intersection of immunology and neuroscience. *J Clin Invest* 122:1146–1148. doi:10.1172/JCI63493

Stephan AH, Barres BA, Stevens B (2012) The complement system: an unexpected role in synaptic pruning during development and disease. *Annu Rev Neurosci* 35:369–389. doi:10.1146/annurev-neuro-061010-113810

Stutman O (1974) Tumor development after 3-methylcholanthrene in immunologically deficient athymic-nude mice. *Science* 183:534–536

Tanaka A, Sakaguchi S (2017) Regulatory T cells in cancer immunotherapy. *Cell Research* 27:109–118. doi:10.1038/cr.2016.151

Tanchot C, Lemonnier FA, Pérarnau B, et al. (1997) Differential requirements for survival and proliferation of CD8 naïve or memory T cells. *Science* 276:2057–2062

Tauber AI (1994) *The Immune Self: Theory or Metaphor?* Cambridge University Press, Cambridge

Tauber AI (1991) *Organism and the Origins of Self*. Kluwer, Dordrecht

Tauber AI (1997) Historical and philosophical perspectives concerning immune cognition. *Journal of the History of Biology* 30:419–440

Thomas L (1959) Discussion. In: Lawrence HS (ed.) *Cellular and Humoral Aspects of the Hypersensitive States*. Hoeber-Harper, New York, pp 529–532

Thomas L (1982) On immunosurveillance in human cancer. *Yale J Biol Med* 55:329–333

van Nood E, Vrieze A, Nieuwdorp M, et al. (2013) Duodenal infusion of donor feces for recurrent *Clostridium difficile*. *New England Journal of Medicine* 368:407–415. doi:10.1056/NEJMoa1205037

Veiga-Fernandes H, Pachnis V (2017) Neuroimmune regulation during intestinal development and homeostasis. *Nat Immunol* 18:116–122. doi:10.1038/ni.3634

Virgin HW (2014) The virome in mammalian physiology and disease. *Cell* 157:142–150. doi:10.1016/j.cell.2014.02.032

von Boehmer H, Kisielow P (1990) Self-nonself discrimination by T cells. *Science* 248:1369–1373. doi:10.1126/science.1972594

Vonderheide RH (2018) The coming immunotherapy revolution is our greatest hope yet for defeating cancer. https://www.telegraph.co.uk/science/2018/07/02/coming-immunotherapy-revolution-greatest-hope-yet-defeating/ *The Telegraph*

Vos T, Allen C, Arora M, et al. (2016) Global, regional, and national incidence, prevalence, and years lived with disability for 310 diseases and injuries, 1990–2015: a systematic analysis for the Global Burden of Disease Study 2015. *The Lancet* 388:1545–1602. doi:10.1016/S0140-6736(16)31678-6

Vuong HE, Yano JM, Fung TC, Hsiao EY (2017) The microbiome and host behavior. *Annu Rev Neurosci* 40:21–49. doi:10.1146/annurev-neuro-072116-031347

Wang H, Naghavi M, Allen C, et al. (2016) Global, regional, and national life expectancy, all-cause mortality, and cause-specific mortality for 249 causes of death, 1980–2015: a systematic analysis for the Global Burden of Disease Study 2015. *The Lancet* 388:1459–1544. doi:10.1016/S0140-6736(16)31012-1

Weiner HL, Frenkel D (2006) Immunology and immunotherapy of Alzheimer's disease. *Nature Reviews Immunology* 6:404–416. doi:10.1038/nri1843

Welch HG, Black WC (2010) Overdiagnosis in cancer. *J Natl Cancer Inst* 102:605–613. doi:10.1093/jnci/djq099

Wiggins D (2016) Activity, process, continuant, substance, organism. *Philosophy* 91:269–280. doi:10.1017/S0031819115000637

Williams S, Sakic B, Hoffman SA (2010) Circulating brain-reactive autoantibodies and behavioral deficits in the MRL model of CNS lupus. *Journal of Neuroimmunology* 218:73–82. doi:10.1016/j.jneuroim.2009.10.008

Wimsatt WC (1972) Complexity and organization. *PSA: Proceedings of the Biennial Meeting of the Philosophy of Science Association* 1972:67–86

Wing K, Sakaguchi S (2010) Regulatory T cells exert checks and balances on self tolerance and autoimmunity. *Nat Immunol* 11:7–13. doi:10.1038/ni.1818

Wolchok JD, Kluger H, Callahan MK, et al. (2013) Nivolumab plus Ipilimumab in advanced melanoma. *New England Journal of Medicine* 369:122–133. doi:10.1056/NEJMoa1302369

Wolfe RA, Roys EC, Merion RM (2010) Trends in organ donation and transplantation in the United States, 1999–2008. *American Journal of Transplantation* 10:961–972. doi:10.1111/j.1600-6143.2010.03021.x

Wright L (1973) Functions. *The Philosophical Review* 82:139–168. doi:10.2307/2183766

Wu Y, Dissing-Olesen L, MacVicar BA, Stevens B (2015) Microglia: dynamic mediators of synapse development and plasticity. *Trends Immunol* 36:605–613. doi:10.1016/j.it.2015.08.008

Wynn TA, Chawla A, Pollard JW (2013) Macrophage biology in development, homeostasis and disease. *Nature* 496:445–455. doi:10.1038/nature12034

Wynn TA, Vannella KM (2016) Macrophages in tissue repair, regeneration, and fibrosis. *Immunity* 44:450–462. doi:10.1016/j.immuni.2016.02.015

Xu J, Gordon JI (2003) Honor thy symbionts. *PNAS* 100:10452–10459. doi:10.1073/pnas.1734063100

Zardavas D, Irrthum A, Swanton C, Piccart M (2015) Clinical management of breast cancer heterogeneity. *Nature Reviews Clinical Oncology* 12:381–394. doi:10.1038/nrclinonc.2015.73

Zilber-Rosenberg I, Rosenberg E (2008) Role of microorganisms in the evolution of animals and plants: the hologenome theory of evolution. *FEMS Microbiol Rev* 32:723–735. doi:10.1111/j.1574-6976.2008.00123.x

Zitvogel L, Ma Y, Raoult D, et al. (2018) The microbiome in cancer immunotherapy: diagnostic tools and therapeutic strategies. *Science* 359:1366–1370. doi:10.1126/science.aar6918

Acknowledgments

I would like to thank my colleague and friend Jean-François Moreau, Professor of Immunology at the University of Bordeaux, for his continuously inspiring "big picture" of immunology. It has been a pleasure and honor to work at his side since 2014. Maël Lemoine and Jean-François Moreau read the entire manuscript and helped me to make it sharper. Lucie Laplane, Jan Pieter Konsman, and Anya Plutynski each read one specific section and made very useful comments. I owe special thanks to Wiebke Bretting, who drew several figures of this Element. William Morgan made very useful suggestions. Many thanks for discussions about immunology and philosophy to Lynn Chiu, Marc Daëron, Louis Du Pasquier, John Dupré, Gérard Eberl, Melinda Fagan, Scott Gilbert, Peter Godfrey-Smith, Deborah Gordon, Paul Griffiths, Matt Haber, Philippe Huneman, Akiko Iwasaki, Jonathan Kipnis, Philippe Kourilsky, Bruno Lemaitre, Tim Lewens, Richard Lewontin, Margaret McFall-Ngai, Alberto Mantovani, Sarkis Mazmanian, Ruslan Medzhitov, Michel Morange, Alvaro Moreno, Samir Okasha, Susan Oyama, Philippe Sansonetti, Ken Schaffner, Elliott Sober, Kim Sterelny, Joan Strassmann, Marie-Elise Truchetet, Skip Virgin, and Eric Vivier. My gratitude and admiration go to Jean Gayon (1949–2018), my supervisor and friend, who is so greatly missed. This project has received funding from the European Research Council (ERC) under the European Union's Horizon 2020 research and innovation program – grant agreement n° 637647 – IDEM. I thank all the past and present members of that project. I also thank all the members of the Conceptual Biology and Medicine group, as well as the ImmunoConcept lab as a whole, the CNRS, and the University of Bordeaux. I am lucky to work in such a friendly and inspiring interdisciplinary environment.

This Element is dedicated to Jean-François Moreau, who embodies immunology.

Cambridge Elements

Philosophy of Biology

Grant Ramsey
KU Leuven

Grant Ramsey is a BOFZAP research professor at the Institute of Philosophy, KU Leuven, Belgium. His work centers on philosophical problems at the foundation of evolutionary biology. He has been awarded the Popper Prize twice for his work in this area. He also publishes in the philosophy of animal behavior, human nature and the moral emotions. He runs the Ramsey Lab (theramseylab.org), a highly collaborative research group focused on issues in the philosophy of the life sciences.

Michael Ruse
Florida State University

Michael Ruse is the Lucyle T. Werkmeister Professor of Philosophy and the Director of the Program in the History and Philosophy of Science at Florida State University. He is Professor Emeritus at the University of Guelph, in Ontario, Canada. He is a former Guggenheim fellow and Gifford lecturer. He is the author or editor of over sixty books, most recently *Darwinism as Religion: What Literature Tells Us about Evolution*; *On Purpose*; *The Problem of War: Darwinism, Christianity, and their Battle to Understand Human Conflict*; and *A Meaning to Life*.

About the Series

This Cambridge Elements series provides concise and structured introductions to all of the central topics in the philosophy of biology. Contributors to the series are cutting-edge researchers who offer balanced, comprehensive coverage of multiple perspectives, while also developing new ideas and arguments from a unique viewpoint.

Cambridge Elements

Philosophy of Biology

Elements in the Series

The Biology of Art
Richard A. Richards

The Darwinian Revolution
Michael Ruse

Ecological Models
Jay Odenbaugh

Mechanisms in Molecular Biology
Tudor M. Baetu

The Role of Mathematics in Evolutionary Theory
Jun Otsuka

Paleoaesthetics and the Practice of Paleontology
Derek D. Turner

Philosophy of Immunology
Thomas Pradeu

A full series listing is available at: www.cambridge.org/EPBY